VOLUME ONE HUNDRED AND NINETY

# ADVANCES IN
# IMAGING AND
# ELECTRON PHYSICS

EDITOR-IN-CHIEF

# Peter W. Hawkes
*CEMES-CNRS*
*Toulouse, France*

VOLUME ONE HUNDRED AND NINETY

# ADVANCES IN
# IMAGING AND
# ELECTRON PHYSICS

Edited by

**PETER W. HAWKES**
*CEMES-CNRS, Toulouse, France*

AMSTERDAM • BOSTON • HEIDELBERG • LONDON
NEW YORK • OXFORD • PARIS • SAN DIEGO
SAN FRANCISCO • SINGAPORE • SYDNEY • TOKYO
Academic Press is an imprint of Elsevier

Cover photo credit:
Grzegorz Wielgoszewski and Teodor Gotszalka; Scanning Thermal Microscopy (SThM):
How to Map Temperature and Thermal Properties at the Nanoscale
Advances in Imaging and Electron Physics (2015) 190, pp. 177–222

Academic Press is an imprint of Elsevier
225 Wyman Street, Waltham, MA 02451, USA
525 B Street, Suite 1800, San Diego, CA 92101-4495, USA
125 London Wall, London, EC2Y 5AS, UK
The Boulevard, Langford Lane, Kidlington, Oxford OX5 1GB, UK

First edition 2015

ISBN: 978-0-12-802380-8
ISSN: 1076-5670

For information on all Academic Press publications
visit our website at http://store.elsevier.com/

Working together
to grow libraries in
developing countries

www.elsevier.com • www.bookaid.org

# CONTENTS

## 4. Scanning Thermal Microscopy (SThM): How to Map Temperature and Thermal Properties at the Nanoscale     177

Grzegorz Wielgoszewski and Teodor Gotszalk

# PREFACE

This volume of the Advances is concerned with various aspects of microscopy: *in situ* and correlative microscopy, the new family of detectors for the electron microscope and scanning thermal microscopy. In addition, I have included a supplement to the list of (electron) microscopy conference proceedings published in volume 127.

The first long chapter, compiled by N. de Jonge, contains extended abstracts of papers presented at a recent meeting on in situ and correlative electron microscopy. The subjects studied range from biology, through biophysics to materials science. This usefully complements the abstracts of the first meeting on this subject, published in an earlier volume (179, 2013, 137–202).

This is followed by an account of the present state of development of direct detectors for cryo-electron microscopy by A.R.Faruqi, R. Henderson and G. McMullan of the MRC Laboratory of Molecular Biology in Cambridge, where so many of the electron microscope techniques used in molecular biology were developed. These new detectors have allowed many hitherto inaccessible observations to be made and I am delighted to publish this authoritative account of the underlying physics and technology here.

The third chapter is a list of the dates and venues of the principal series of congresses on (electron) microscopy as well as several meetings in related areas, notably charged-particle optics. This is much less ambitious than its predecessor (AIEP 127, 2003, 207–379), where full details of many national meetings were also listed.

The volume ends with a fascinating account of scanning thermal microscopy by G. Wielgoszewski and T. Gotszalk. This forms a concise monograph on the subject for the authors cover the history and principles of scanning thermal microscopy, thermal probes, measurement devices, metrology and a few applications. An opening section describes how this form of microscopy developed from the original scanning probe microscope.

I thank the authors on behalf of all our readers for taking trouble to make their material accessible to a wide readership.

PETER W. HAWKES

# FUTURE CONTRIBUTIONS

**H.-W. Ackermann**
Electron micrograph quality

**S. Ando**
Gradient operators and edge and corner detection

**J. Angulo**
Mathematical morphology for complex and quaternion-valued images

**D. Batchelor**
Soft x-ray microscopy

**E. Bayro Corrochano**
Quaternion wavelet transforms

**C. Beeli**
Structure and microscopy of quasicrystals

**M. Berz and K. Makino** Eds (Vol. 191)
Femtosecond electron imaging and spectroscopy

**C. Bobisch and R. Möller**
Ballistic electron microscopy

**F. Bociort**
Saddle-point methods in lens design

**K. Bredies**
Diffusion tensor imaging

**A. Broers**
A retrospective

**R.E. Burge** (Vol. 191)
A scientific autobiography

**N. Chandra and R. Ghosh**
Quantum entanglement in electron optics

**A. Cornejo Rodriguez and F. Granados Agustin**
Ronchigram quantification

**L.D. Duffy and A. Dragt**
Eigen-emittance

**J. Elorza**
Fuzzy operators

**M. Ferroni**
Transmission microscopy in the scanning electron microscope

**R.G. Forbes**
Liquid metal ion sources

**P.L. Gai and E.D. Boyes**
Aberration-corrected environmental microscopy

**V.S. Gurov, A.O. Saulebekov and A.A. Trubitsyn**
Analytical, approximate analytical and numerical methods for the design of energy analyzers

**M. Haschke**
Micro-XRF excitation in the scanning electron microscope

**R. Herring and B. McMorran**
Electron vortex beams

**M.S. Isaacson**
Early STEM development

**K. Ishizuka**
Contrast transfer and crystal images

**K. Jensen, D. Shiffler and J. Luginsland**
Physics of field emission cold cathodes

**M. Jourlin**
Logarithmic image processing, the LIP model. Theory and applications

**U. Kaiser**
The sub-Ångström low-voltage electron microcope project (SALVE)

**C.T. Koch**
In-line electron holography

**O.L. Krivanek**
Aberration-corrected STEM

**M. Kroupa**
The Timepix detector and its applications

**B. Lencová**
Modern developments in electron optical calculations

**H. Lichte**
Developments in electron holography

**M. Matsuya**
Calculation of aberration coefficients using Lie algebra

**J.A. Monsoriu**
Fractal zone plates

**L. Muray**
Miniature electron optics and applications

**M.A. O'Keefe**
Electron image simulation

**V. Ortalan**
Ultrafast electron microscopy

**D. Paganin, T. Gureyev and K. Pavlov**
Intensity-linear methods in inverse imaging

**N. Papamarkos and A. Kesidis**
The inverse Hough transform

**Q. Ramasse and R. Brydson**
The SuperSTEM laboratory

**B. Rieger and A.J. Koster**
Image formation in cryo-electron microscopy

**P. Rocca and M. Donelli**
Imaging of dielectric objects

**J. Rodenburg**
Lensless imaging

**J. Rouse, H.-n. Liu and E. Munro**
The role of differential algebra in electron optics

**J. Sánchez**
Fisher vector encoding for the classification of natural images

**P. Santi**
Light sheet fluorescence microscopy

**R. Shimizu, T. Ikuta and Y. Takai**
Defocus image modulation processing in real time

**T. Soma**
Focus-deflection systems and their applications

**I.F. Spivak–Lavrov**
Analytical methods of calculation and simulation of new schemes of static and time-of-flight mass spectrometers

**J. Valdés**
Recent developments concerning the Système International (SI)

# CONTRIBUTORS

**Niels de Jonge**
INM-Leibniz Institute for New Materials, Campus D2 2, and Department of Physics, University of Saarland, Campus A5 1, 66123 Saarbrücken, Germany

**A.R. Faruqi**
MRC Laboratory of Molecular Biology, Francis Crick Ave., Cambridge Biomedical Campus, Cambridge CB2 0QH, UK

**Teodor Gotszalk**
Wrocław University of Technology, Faculty of Microsystem Electronics and Photonics, Nanometrology Group, ul. Z. Janiszewskiego 11/17, PL-50372 Wrocław, Poland

**Peter W. Hawkes**
CEMES-CNRS, Toulouse

**Richard Henderson**
MRC Laboratory of Molecular Biology, Francis Crick Ave., Cambridge Biomedical Campus, Cambridge CB2 0QH, UK

**Greg McMullan**
MRC Laboratory of Molecular Biology, Francis Crick Ave., Cambridge Biomedical Campus, Cambridge CB2 0QH, UK

**Grzegorz Wielgoszewski**
Wrocław University of Technology, Faculty of Microsystem Electronics and Photonics, Nanometrology Group, ul. Z. Janiszewskiego 11/17, PL-50372 Wrocław, Poland, and University College Dublin, School of Physics, Science Centre North, Belfield, Dublin 4, Ireland

CHAPTER ONE

# CISCEM 2014
## Proceedings of the Second Conference on In situ and Correlative Electron Microscopy, Saarbrücken, Germany, October 14–15, 2014

## Niels de Jonge[a,b,]*
[a]INM-Leibniz Institute for New Materials, Campus D2 2, 66123 Saarbrücken, Germany
[b]Department of Physics, University of Saarland, Campus A5 1, 66123 Saarbrücken, Germany
*Corresponding author: e-mail address: niels.dejonge@inm-gmbh.de

## PREFACE

One of the key challenges at the forefront of today's electron micros-copy research is to observe processes at the nanoscale under relevant condi-tions. For samples from the materials science, this is accomplished by in situ electron microscopy. Movies—even at atomic resolution—are recorded of processes at high temperatures, in gaseous environments, or in liquids, while carefully taking into account the effect of the electron beam. For most bio-logical samples, the electron beam impact prevents acquiring the time-lapse date, and research is mostly directed toward correlative light- and electron microscopy often using proteins labels. Single electron microscopic images are preferentially recorded in amorphous ice, or liquid. A conference dis-cussing these topics was held for the second time at the INM-Leibniz Insti-tute for New Materials on October 14–15, 2014, in Saarbrücken, Germany. The conference on in situ and correlative electron microscopy CISCEM 2014 aimed to bring together an interdisciplinary group of scientists from the fields of biology, materials science, chemistry, and physics to discuss future directions of electron microscopy research. The venue was the Aula at Saarland University.

The conference opened with a session on correlative and in situ electron microscopy in biology. Keynote speaker Wolfgang Baumeister gave a broad overview of in situ transmission electron microscopy (TEM) of proteins and cells embedded in amorphous ice. Recent advances in correlative light and electron microscopy were discussed by the invited speakers, including Ben Giepmans and Paul Verkade. Deborah F. Kelly and Diana B. Peckys

*Advances in Imaging and Electron Physics*, Volume 190
ISSN 1076-5670
http://dx.doi.org/10.1016/bs.aiep.2015.02.004

1

presented the topic of electron microscopy of cells and viruses in liquid. The second session involved in situ observations of biomineralization processes. A highlight on this topic was a presentation by James de Yoreo, showing movies of such processes with atomic resolution. Bio-degradation processes were studied in situ by Damien Alloyeau.

The first day of the conference also accommodated a session with (non-scientific) corporate presentations (not reflected in this chapter). This day ended with a poster session including a total of 29 posters on the following topics, movement of nanoparticles, designing in situ experiments, high-temperature and other in situ experiments, experiments in biology, and experiments in metals.

The second day started with a dedicated session discussing various aspects of the design of in situ experiments. The two invited speakers at this session were Eva Olsson and Patrica Abellan. The most important topic was experiments in a liquid environment. High-temperature and other experiments were presented in the fourth session, including also a presentation on in situ characterization of battery materials. The conference concluded with a session on in situ TEM of catalytic nanoparticles in gaseous environments. Invited speakers were Renu Sharma and Jakob Wagner.

This chapter contains a selection of the extended abstracts submitted for the conference.

**Niels de Jonge**
**December 19, 2014**

## LIST OF CONTRIBUTIONS

### Session 1: Correlative and In Situ Electron Microscopy in Biology

- Keynote Lecture: "Electron Cryomicroscopy Ex Situ and In Situ," Wolfgang Baumeister
- Invited: "Correlative Light and Electron Microscopy (CLEM): Ultra-structure Lights Up," Ben Giepmans
- Invited: "Correlative Light Electron Microscopy $1+1=3$," Paul Verkade
- Invited: "Improving Our Vision of Nanobiology with In Situ TEM," Deborah F. Kelly

- "Environmental Scanning Electron Microscopy for Studying Proteins and Organelles in Whole, Hydrated Eukaryotic Cells with Nanometer Resolution," Diana B. Peckys
- "Integrated CLEM—Still Bridging the Resolution Gap," Gerhard A. Blab
- "Cellular Membrane Rearrangements Induced by Hepatitis C Virus," Inés Romero-Brey

## Session 2: In Situ Observations of Biomineralization Processes

- Invited: "Nucleation and Particle Mediated Growth in Mineral Systems Investigated by Liquid-Phase TEM," James de Yoreo
- Invited: "Studying the In Situ Growth and Biodegradation of Inorganic Nanoparticles by Liquid-Cell Aberration Corrected TEM," Damien Alloyeau
- "In Situ TEM Shows Ion Binding Is Key to Directing CaCO3 Nucleation in a Biomimetic Matrix," Paul Smeets
- "Crystallisation of Calcium Carbonate Studied by Liquid Cell Scanning Transmission Electron Microscopy," Andreas Verch

## Session 3: Designing In Situ Experiments

- Invited: "Studies of Transport Properties using In Situ Microscopy," Eva Olsson
- Invited: "Calibrated In Situ Transmission Electron Microscopy for the Study of Nanoscale Processes in Liquids," Patricia Abellan
- "Microchip-Systems for In Situ Electron Microscopy of Processes in Gases and Liquids," Kristian Mølhave
- "Scanning Transmission Electron Microscopy of Liquid Specimens," Niels de Jonge
- "Scanning Electron Spectro-Microscopy in Liquids and Dense Gaseous Environment through Electron Transparent Graphene Membranes," Andrei Kolmakov

## Session 4: High-Temperature and Other In Situ Experiments

- "In Situ HT-ESEM Observation of CeO2 Nanospheres Sintering: From Neck Elaboration to Microstructure Design," Galy I. Nkou Bouala
- "In Situ Transmission Electron Microscopy of High-Temperature Phase Transitions in Ge-Sb-Te Alloys," Katja Berlin

## Session 5: In Situ Tem of Catalytic Nanoparticles

- Invited: "Correlative Microscopy for In Situ Characterization of Catalyst Nanoparticles Under Reactive Environments," Renu Sharma
- Invited: "Applications of Environmental TEM for Catalysis Research," Jakob B. Wagner

 **SELECTED POSTERS**

- "Gold Nanoparticle Movement in Liquid Investigated by Scanning Transmission Electron Microscopy," Marina Pfaff & Niels de Jonge
- "Correlating Scattering and Imaging Techniques: In Situ Characterization of Au Nanoparticles Using Conventional TEM," Dimitri Vanhecke et al.
- "The Effects of Salt Concentrations and pH on the Stability of Gold Nanoparticles in Liquid Cell STEM Experiments," Andreas Verch et al.
- "Bridging the Gap Between Electrochemistry and Microscopy: Electrochemical IL-TEM and In Situ Electrochemical TEM Study," Nejc Hodnik et al.
- "Using a Combined TEM/Fluorescence Microscope to Investigate Electron Beam–Induced Effects on Fluorescent Dyes Mixed into an Ionic Liquid," Eric Jensen et al.
- "Microfabricated Low-Thermal Mass Chips System for Ultra-Fast Temperature Recording During Plunge freezing for Cryofixation," Simone Laganá et al.
- "In Situ SEM Cell for Analysis of Electroplating and Dissolution of Cu," Rolf Møller-Nilsen et al.
- "Integrated Correlative Light and Electron Microscopy (iCLEM) for Optical Sectioning of Cells Under Vacuum and Near-Native Conditions to Investigate Membrane Receptors," Josey Sueters et al.
- "In Situ Dynamic ESEM Observations of Basic Groups of Parasites," Š. Mašová et al.
- "Determination of Nitrogen Gas Pressure in Hollow Nanospheres Produced by Pulsed Laser Deposition in Ambient Atmosphere by Combined HAADF-STEM and Time-Resolved EELS Analysis," Sašo Šturm
- "Platelet granule secretion: A (Cryo)-Correlative Light and Electron Microscopy Study," K. Engbers-Moscicka et al.

# Correlative and In Situ Electron Microscopy in Biology

# Electron Cryomicroscopy Ex Situ and In Situ

**Wolfgang Baumeister***

Max-Planck-Institute of Biochemistry, Am Klopferspitz 18, D-82152 Martinsried, Germany
*Corresponding author: e-mail address: baumeist@biochem.mpg.de

Today, there are three categories of biomolecular electron microscopy (EM): (1) electron crystallography, (2) single-particle analysis and (3) electron tomography. Ideally, all three imaging modalities are applied to frozen-hydrated samples, ensuring that they are studied in the most lifelike state that is physically possible to achieve. Vitrified aqueous samples are very radiation sensitive and consequently, cryo-EM images must be recorded at minimal electron beam exposures, limiting their signal-to-noise ratio. Therefore, the high-resolution information of images of unstained and vitrified samples must be retrieved by averaging-based noise reduction, which requires the presence of repetitive structure. Averaging can obviously not be applied to pleomorphic structures such as organelles and cells (Fitting Kourkoutis et al., 2012; Leis et al., 2009).

Electron crystallography requires the existence of two-dimensional crystals, natural or synthetic, and averaging is straightforward given the periodic arrangement of the molecules under scrutiny. In principle, electron crystallography is a high-resolution technique as demonstrated successfully with a number of structures, particularly membrane proteins. Often, however, the same structures can be studied by X-ray crystallography, which tends to be faster and can attain atomic resolution more easily.

In contrast, EM single particle analysis (arguably a misnomer since it involves the averaging over large numbers of identical particles) has become one of the pillars of modern structural biology. The amount of material needed is minute, and some degree of heterogeneity, compositional or conformational, is tolerable since image classification can be used for further purification in silico. It is particularly successful in structural studies of very large macromolecular complexes where the traditional methods often fail. In principle, single-particle analysis can attain near-atomic resolution, but in practice, this often remains an elusive goal. This is changing, however, with the advent of new technology, particularly detectors with improved performance. But even intermediate resolution (subnanometer) structures of very large complexes can provide an excellent basis for hybrid or integrative approaches in which high-resolution structures of components and orthogonal data, such as distance restraints, are used to generate atomic models.

Electron cryotomography can be used to study the three-dimensional organization of nonrepetitive objects (Lucic *et al.*, 2005). Most cellular structures fall into this category. In order to obtain three-dimensional reconstructions of objects with unique topologies, it is necessary to acquire data sets with different angular orientations of the sample by physical tilting. The challenge is to obtain large numbers of projections covering as wide a tilt range as possible and, at the same time, to minimize the cumulative electron dose. This is achieved by means of elaborate automated acquisition procedures. Electron cryotomography can provide medium-resolution, three-dimensional images of a wide range of biological structures from isolated supramolecular assemblies to organelles and whole cells. It allows the visualization of molecular machines in their unperturbed functional environments (in situ structural biology) and ultimately the mapping of entire molecular landscapes (visual proteomics; Robinson *et al.*, 2007).

Until recently, the use of electron cryotomography was restricted to relatively thin samples, such as prokaryotic cells or the margins of eukaryotic cells. This has changed with the advent of focused-ion beam (FIB) micromachining and developments that allowed the application of this technology to samples embedded in vitreous ice. This allows the cutting of "windows" providing views of the interior of thicker samples such as eukaryotic cells. By combining the FIB with correlative fluorescence microscopy, it is now possible to navigate large cellular landscapes and to select and target specific areas of interest (Villa *et al.*, 2013).

Given the full signal-to-noise ratio of the tomograms, it can be challenging to interpret them and take advantage of their rich information content. Image denoising can improve the signal-to-noise ratio by reducing the noise while preserving the features of interest. Segmentation separates the structures of interest from the background and allows their three-dimensional visualization and quantitative analysis. Larger molecular structures can be identified in tomograms by pattern recognition methods using a template structure, and once their location and orientation is determined, identical structures can be extracted computationally and averaged. Therefore, electron tomography has unique potential to *bridge* the divide between molecular and cellular structural studies, perhaps the most exciting frontier in structural biology (Villa *et al.*, 2013).

# REFERENCES

Fitting, Kourkoutis, L., Plitzko, J.M., & Baumeister W. (2012). Electron microscopy of biological materials at the nanometer scale. *Annual Review of Materials Science, 42*, 33–58.

Leis, A., Rockel B., Andrees, L., & Baumeister W. (2009). Visualizing cells at the nanoscale. *Trends in Biochemical Sciences, 34*, 60–70.

Lucic, V., Förster, F., & Baumeister W. (2005). Structural studies by electron tomography: From cells to molecules. *Annual Review of Biochemistry, 74*, 833–865.

Robinson, C.V., Sali, A., & Baumeister W. (2007). The molecular sociology of the cell. *Nature 450*, 973–982.

Villa, E., Schaffer, M., Plitzko, J.M., & Baumeister, W. (2013). Opening windows into the cell: Focused-ion-beam milling for cryo-electron tomography. *Current Opinion in Structural Biology. 23*, 1–7.

# Correlative Light and Electron Microscopy (CLEM)

**Ben N.G. Giepmans***

Department of Cell Biology, University of Groningen, University Medical Center Groningen, The Netherlands

*Corresponding author: e-mail address: b.n.g.giepmans@umcg.nl; www.cellbiology.nl

Today, I will first focus on recent developed labeling strategies for probes that allow Correlated light and electron microscopy (CLEM) (Giepmans, 2008; Sjollema *et al.*, 2012). These include particles (gold, quantum dots) to highlight endogenous proteins, but also genetically encoded probes, as well as traditionally used stains for light microscopy (LM) that aid in electron microscopy (EM)–analysis of samples. Probes that can be detected only in a single modality and require image overlay, as well as combinatorial probes that can be visualized both at LM and EM will be discussed. In addition, published (Ravelli *et al.*, 2013) and new approaches for large-scale EM to visualize macromolecules and organelles in the context of organized cell systems and tissues will be covered (www.nanotomy.nl). Matching the areas of acquisition in CLEM and EM will not only increase understanding of the molecules in the context, but also is a straightforward manner to combine the LM and EM image data. Covering a wide variety of probes and approaches for image overlay will help to enable (new) users to implement CLEM to better understand how molecules (mal)function in biology.

 ## LARGE-SCALE EM ("NANOTOMY") ALLOWS ANALYSIS OF TISSUE UP TO THE MOLECULAR LEVEL

Figure 1 snapshots taken from www.nanotomy.nl, an open-source database, zooming into the boxed areas (clockwise). Note that the islet of Langerhans is identifiable (left), but also cells and organelles, as well as macromolecular complexes (ribosomes, nuclear pores, etc). Data from Ravelli *et al.* (2013).

See Ravelli *et al.* (2013) and click www.nanotomy.nl for details.

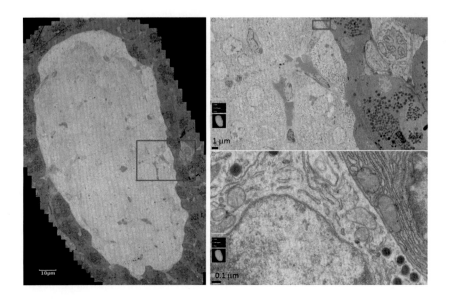

## REFERENCES

Giepmans, B.N. (2008). Bridging fluorescence microscopy and electron microscopy. *Histochemistry and Cell Biology*, *130*(2), 211–217.

Ravelli, R.B.G., Kalicharan, R.D., Avramut, C.M., Sjollema, K.A., Pronk, J.W., Dijk, F., et al. (2013). Destruction of tissue, cells and organelles in type 1 diabetic rats presented at macromolecular resolution. *Scientific Reports, 3*, 1804; doi:10.1038/srep01804.

Sjollema, K.A., Schnell, U., Kuipers, J., Kalicharan, R., & Giepmans, B.N.G. (2012). Correlated light microscopy and electron microscopy. *Methods in Cell Biology, 111,* 157–173.

# Correlative Light Electron Microscopy, 1 + 1 = 3

**Lorna Hodgson, Paul Verkade\***

Wolfson Bioimaging Facility, Schools of Biochemistry and Physiology & Pharmacology, Medical Sciences Building, University Walk, University of Bristol, Bristol, UK

\*Corresponding author: e-mail address: p.verkade@bristol.ac.uk

Correlative light electron microscopy combines the strengths of light and electron microscopy in one experiment, and the sum total of such an experiment should provide more data/insight than each technique alone (hence 1 + 1 = 3). There are many ways to perform a CLEM experiment, and a variety of microscopy modalities can be combined. The choice of these instruments should primarily depend on the scientific question to be answered.

A CLEM experiment can usually be divided into three parts; probes, processing, and analysis. I will discuss three processing techniques based on light microscopy in conjunction with transmission electron microscopy, each with its advantages and challenges.

The first is based on the use of coverslips with a finder pattern, it allows live cell imaging and captures an event of interest using chemical fixation (Figure 1; Hodgson *et al.*, 2014a), A second uses the Tokuyasu cryo immuno

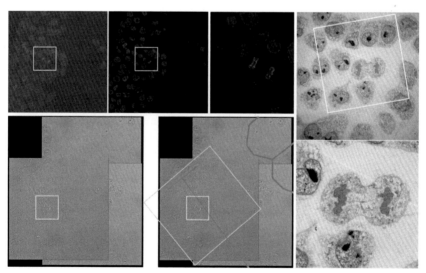

**Figure 1** CLEM allows the identification and subsequent high-resolution analysis of 1 special cell among hundreds. In the upper-left row, a dividing cell (1 in >100) is identified based on its DNA staining (blue). With the aid of embossed coverslips (for full details, see Hodgson *et al.*, 2014a) the dividing cell can be traced back in the EM and studied at higher resolutions. (See the color plate.)

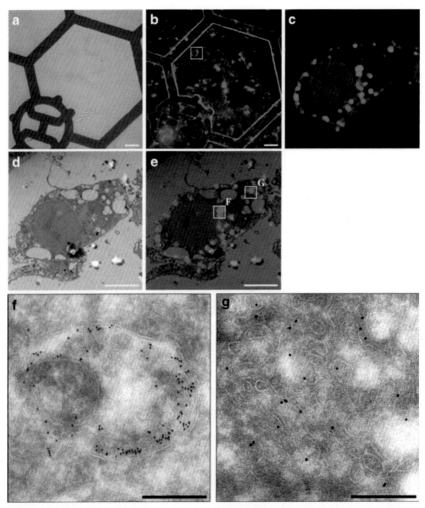

**Figure 2** CLEM using Tokuyasu cryo-immuno gold labeling, an excellent way to zoom into specific structures with high labeling efficiency. *Reproduced from Hodgson, Tavaré, & Verkade (2014).*

labelling to trace back objects of interest (Figure 2; Hodgson *et al.*, 2014b), this allows for relatively high immunolabeling efficiencies but is almost impossible in combination with live cell imaging. The third is based on cryofixation to obtain the best possible preservation of ultrastructure (Verkade, 2008; Brown *et al.*, 2012). This allows us to capture events that would be lost because of chemical fixation (e.g., membrane tubules). It allows for live cell imaging, but immunolabeling options are limited.

# REFERENCES

Brown, E., Van Weering, J., Sharp, T., Mantell, J., & Verkade, P. (2012). Capturing endocytic segregation events with HPF-CLEM. *Methods in Cell Biology, 111: Correlative Light and Electron Microscopy,* 175–201.

Hodgson, L, Nam, D., Mantell, J., Achim A., & Verkade, P. (2014a). Retracing in correlative light electron microscopy: Where is my object of interest? *Methods in Cell Biology, 124: Correlative Light and Electron Microscopy II,* 1–21.

Hodgson, L, Tavaré, J., & Verkade, P. (2014b). Development of a quantitative correlative light electron microscopy technique to study GLUT4 trafficking. *Protoplasma, 251,* 403–416.

Verkade, P. (2008). Moving EM: The rapid transfer system as a new tool for correlative light and electron microscopy and high throughput for high-pressure freezing. *Journal of Microscopy. 230,* 317–328.

# Improving Our Vision of Nanovirology with In Situ TEM

Andrew C. Demmert[a], Madeline J. Dukes[b], Sarah M. McDonald[a], Deborah F. Kelly[a],*

[a]Virginia Tech Carilion School of Medicine and Research Institute, Virginia Tech Roanoke, VA 24016
[b]Application Science Division, Protochips, Inc., Raleigh NC 27606
*Corresponding author: e-mail address: debkelly@vt.edu

Understanding the fundamental properties of macromolecules has enhanced the development of emerging technologies used to improve biomedical research. Currently, there remains a critical need for innovative platforms that can illuminate the function of biological objects in a native liquid environment. To address this need, we have developed an in situ TEM approach to visualize the dynamic behavior of biomedically relevant macromolecules at the nanoscale. Newly designed silicon nitride-based devices containing integrated microwells were used to enclose active macromolecular specimens in liquid for TEM imaging purposes (Figures 3A, B). With each specimen tested, the integrated microwells could adequately maintain macromolecules in discrete local environments (Dukes et al., 2014) while enabling thin liquid layers to be produced for high-resolution imaging purposes as previously exemplified using gold nanorods (Dukes et al., 2013). This success permitted us to utilize the integrated microwell-designed microchips to examine actively transcribing rotavirus assemblies having native contrast (Dukes et al., 2014).

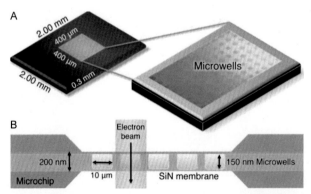

**Figure 3** Next-generation SiN microchips. (A) A schematic to specify the dimensions of the microwell chips used to form the liquid chamber that is positioned with respect to the electron beam **(B).** *Illustrations adapted and reprinted with permission (Dukes et al., 2014).* (See the color plate.)

In developing biochemical experiments to assess viral attributes, we first needed to manufacture competent viral specimens. To accomplish this objective, we purified simian rotavirus double-layered particles (DLPs) (strain SA11-4 F) from monkey kidney MA104 cells, as previously described (Dukes *et al.*, 2014). The proteins that comprised the purified DLPs were analyzed using SDS-PAGE and silver staining (Figure 4A). We found that

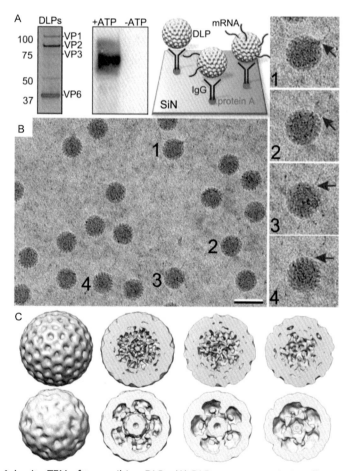

**Figure 4** In situ TEM of transcribing DLPs. (A) DLPs were transcriptionally active upon the addition of ATP to produce [$^{32}$P]-labeled mRNA transcripts. Active DLPs were tethered to SiN microchips coated with Ni-NTA and protein A/IgG adaptors. (B) Images of transcribing DLPs in liquid reveal single-strand mRNA emerging from the viral capsids (1 – 4). Scale bar is 100 nm. (C) 3D structures of active DLPs show movements in their interior during RNA synthesis. *Panels B and C are adapted and reprinted with permission (Dukes et al., 2014).*

DLPs produced in the MA104 cells contained four proteins (VP1, VP2, VP3, and VP6) and that VP4 and VP7 were absent from the formed particles. To verify that our purified DLPs could transcribe viral RNAs, we utilized an in vitro messenger RNA (mRNA) synthesis assay. Each reaction mixture contained DLPs, each NTP, and $[^{32}P]$-UTP, and they were allowed to incubate for 30 min at 37 °C (Figure 4A, +ATP). Negative control reactions also contained each transcription cocktail component except ATP (Figure 4A, −ATP). Radiolabeled mRNA products were detected in the reaction mixtures containing a complete transcription cocktail, and no radiolabeled products were detected in the reaction mixtures lacking ATP. Therefore, this functional analysis confirmed that the purified DLPs used for subsequent imaging analysis were enzymatically active.

We attempted to visualize transcribing rotavirus DLPs using reaction mixtures that were prepared as described previously. An aliquot of each transcription cocktail was added onto Ni-NTA-coated SiN microchips that were previously decorated with His-tagged protein A and IgG polyclonal antibodies against the VP6 capsid protein (Degen et al., 2012; Gilmore et al., 2013) (Figure 4A, schematic). The fluidic microchamber that contained the antibody-bound transcribing DLPs was assembled into the Poseidon specimen holder. Active DLPs were examined using a FEI Spirit Bio-Twin TEM equipped with a $LaB_6$ filament and operating at 120 kV. Images of transcribing DLPs were recorded using an Eagle 2 k HS charge-coupled device (CCD) camera under low-dose conditions (approximately 0.5 electrons/$Å^2$) to minimize beam damage to the viral specimens. The resulting images revealed dynamic attributes of RV pathogens in liquid at 3-nm resolution (Figure 4B). We could also distinguish discrete strands emerging from numerous DLPs. These strands had characteristic shapes and dimensions consistent with being single-stranded viral mRNA transcripts (Figure 4B, 1–4, right panels). No strands were identified in images of our negative control transcription reactions that lacked ATP. We could subsequently use the RELION software package to compute 3D reconstructions of the active DLPs from a single image (Figure 4C). The interiors of the DLP cores revealed movements indicative of protein rearrangements during mRNA synthesis.

## REFERENCES

Degen, K. Dukes, M., Tanner, J.R., & Kelly, D.F. (2012). The development of affinity capture devices—A nanoscale purification platform for biological in situ transmission electron microscopy. *RSC Advances*, 2408–2412.

Dukes, M.J., Thomas, R., Damiano, J., Klein, K.L., Balasubramaniam, S., Kayandan, S., et al. (2014). Improved microchip design and application for in situ transmission electron microscopy of macromolecules. *Microscopy and Microanalysis,* 338–345.

Dukes, M.J., Jacobs, B.W., Morgan, D.G., Hegde, H., & Kelly, D.F. (2013). Visualizing nanoparticle mobility in liquid at atomic resolution. *Chemical Communications,* 3007–3009.

Gilmore, B.L., Showalter, S., Dukes, M.J., Tanner, J.R., Demmert, A.C., McDonald, S.M., & Kelly, D.F. (2013). Visualizing viral assemblies in a nanoscale biosphere. *Lab on a Chip,* 216–219.

# Environmental Scanning Electron Microscopy for Studying Proteins and Organelles in Whole, Hydrated Eukaryotic Cells with Nanometer Resolution

**Diana B. Peckys[a],\*, Niels de Jonge[a,b,c]**
[a]INM-Leibniz Institute for New Materials, Saarbrücken, Germany
[b]Vanderbilt University School of Medicine, Nashville, TN
[c]Physics Department, Saarland University, Saarbrücken, Germany
*Corresponding author: e-mail address: diana.peckys@inm-gmbh.de

The spatial distribution of internalized nanoparticles (NPs), and of membrane proteins tagged with NP labels were studied by imaging whole and hydrated cells with an environmental scanning microscope (ESEM), equipped with a scanning transmission electron microscope (STEM) detector (Figure 5). COS7 fibroblast, A549 lung cancer, and SKBR3 breast cancer cells were grown on silicon microchips with silicon nitride (SiN) membrane windows. One ESEM-STEM study followed the fate of gold nanoparticles (AuNPs) within cells, an important topic in view of the high potential of AuNPs for medical applications. Here, A549 cells took up serum protein coated AuNPs of 10-, 15-, or 30-nm diameters. One or two days after the AuNP uptake cells were fixed and investigated with ESEM-STEM

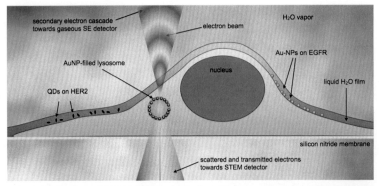

**Figure 5** A schematic of ESEM-STEM of whole cells in a wet state. A focused electron beam (30 keV) is scanned over the fixed and hydrated cells. Contrast is obtained on QDs labeling EGFR or HER2 receptors or on internalized AuNPs. A gaseous secondary electron detector, located above the sample, and a STEM detector, located beneath the sample, simultaneously collect the signals.

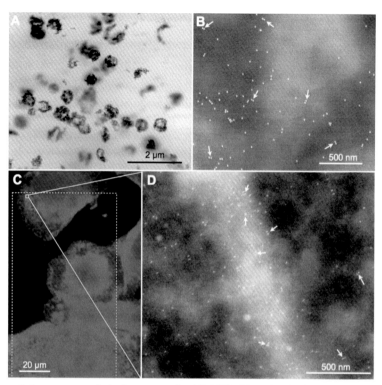

**Figure 6** Images recorded of cancer cells. (A) ESEM-STEM bright field image of A549 lung cancer cells recorded 24 h after the uptake of protein-coated, 30-nm AuNPs. The AuNPs were found in a lining distribution inside lysosomes, with an intracellularly scattered distribution. (B) ESEM-STEM dark field image of A549 showing 12-nm AuNPs specifically bound to EGFRs. A fraction of the receptors appeared as pairs (see examples marked by arrows). (C) Fluorescence image of SKBR3 breast cancer cells with QD-labeled HER2 receptors. The dashed line indicates the borders of the SiN membrane window on the microchip. (D) ESEM-STEM image recorded at the location of the small rectangle in (C). Denser labeling appears on the brighter cellular background interpreted as membrane ruffle. Several dimers of HER2 receptors can be distinguished due to their spatial proximity (examples are marked by arrows).

(Figure 6A), AuNPs were found in a distinct lining pattern within intracellularly scattered lysosomes. The dimensions of 1,106 AuNP-storing lysosomes were determined from 145 whole cells, within a total time (including imaging and analysis) of only 80 h. This study revealed a statistically relevant enlargement effect on the size of the lysosomes of the 30-nm AuNP compared to the smaller NPs (Peckys *et al.*, 2014).

Studied membrane proteins included the epidermal growth factor receptor (EGFR), and the related receptor tyrosine kinase HER2. The mapping of monomers and dimers of these two receptors is important for basic research, as well as for the study of the molecular mechanisms involved in certain anti-cancer drugs. Membrane-bound EGFR or HER2 on live cells were labeled with probes consisting of small protein ligands and AuNPs or fluorescent quantum dots (QDs). After fixation, EGF-AuNP labeled cells were examined directly with ESEM-STEM (Peckys *et al.*, 2013) (Figure 6B), whereas cells labeled with QD probes were studied with correlative fluorescence microscopy and ESEM-STEM (Figure 2C and D). In all cell lines, significant fractions of the labeled receptors appeared as dimers and in small clusters. Hundreds of ESEM-STEM images were recorded of several tens of cells providing nanometer-scaled data from thousands of labels, which were analyzed and quantified by automated image software algorithm. In addition, correlative microscopy confirmed the heterogeneity of nonisogenic cancer cells, manifesting in large variations of EGFR and HER2 expression and distinct spatial distributions on the cell membrane.

In conclusion, ESEM-STEM is an exciting EM methodology for analytic studies of whole cells in their hydrated state with nanometer resolution and in very short timeframes.

## ACKNOWLEDGMENTS

We thank M. Koch for help with the experiments, A. Kraegeloh for support of the experiments, and Protochips Inc, NC, for providing the microchips. We thank E. Arzt for his support through INM. Research in part supported by the Leibniz Competition 2014.

## REFERENCES

Peckys, D.B., Baudoin, J.P., Eder, M., Werner, U., & de Jonge, N. (2013). Epidermal growth factor receptor subunit locations determined in hydrated cells with environmental scanning electron microscopy. *Scientific Reports, 3,* 2626.

Peckys, D.B., & de Jonge, N. (2014). Gold nanoparticle uptake in whole cells in liquid examined by environmental scanning electron microscopy. *Microscopy and Microanalysis, 20*(1), 189–197.

# Integrated CLEM—Still Bridging the Resolution Gap

**G.A. Blab[a,*], M.A. Karreman[a,b], A.V. Agronskaia[a], H.C. Gerritsen[a]**

[a]Molecular Biophysics, Department of Physics, Utrecht University, Postbus 80'000, 3508 TA Utrecht, the Netherlands
[b]EMBL Heidelberg, Team Schwab, Meyerhofstraße 1, 69117 Heidelberg, Germany
*Corresponding author: e-mail address: g.a.blab@uu.nl

While electron microscopy undoubtedly provides unrivalled resolution, localizing relevant parts in a large sample can prove to be prohibitively time consuming. In the past, we have found a workable solution to this problem by the direct integration of a fluorescence scanning microscope inside a TEM column (iLEM, FEI), see Figures 7 and 8. Despite initial challenges

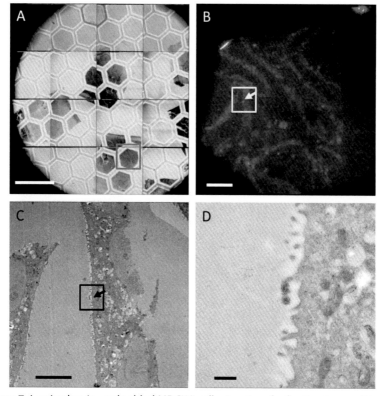

**Figure 7** Lowicryl resin–embedded MDCK II cells. A series of reflection images (A; scale bar 250 $\mu$m) allows us to localize the sample. Immuno-fluorescence (B; scale bar 25 $\mu$m) indicates the locations of acetylated alpha-tubulin as found in cilia. Finally, we obtain TEM images (C; scale bar 5 $\mu$m) and (D; scale bar 500 nm) of the regions of interest found by fluorescence. (See the color plate.)

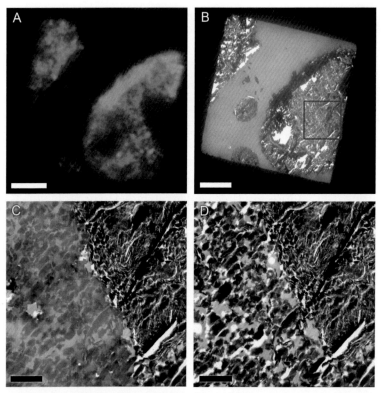

**Figure 8** iLEM analysis of a fluid catalyst cracking (FCC) particle. Active regions in the particle generate a fluorescent product (A; scale bar 10 $\mu$m). A zoom, indicated by a blue box, into the TEM image (B; scale bar 10 $\mu$m) shows that fluorescence and specific morphology are correlated (C, D; scale bar 2 $\mu$m). (See the color plate.)

to combine the two techniques in one instrument, we are now routinely able to register and—using fluorescence probes—accurately localize regions of interest anywhere on a standard EM grid while using most types of TEM sample preparations. We have also shown that we can use intrinsic fluorescence to provide complementary information that is not accessible by EM alone. However, light and electron microscopy remain vastly different methods, with an accordingly large gap in the relevant length scales. In order to bridge this gap, we have recently begun to combine super-resolution light microscopy with TEM.

# REFERENCES

Agronskaia, A.V., Valentijn, J.A., van Driel, J.A., Schneijdenberg, C.T.W.M., Humbel, B.M., van Bergen en Henegouwen, P.M.P., et al. (2008). Integrated fluorescence and transmission electron microscopy: a novel approach to correlative microscopy. *Journal of Structural Biology*, *164*(2) 183–189.

Karreman, M.A., Buurmans, I.L.C., Geus, J.W., Agronskaia, A.V., Ruiz-Martínez, J, Gerritsen, H.C., & Weckhuysen, B.M. (2012). Integrated laser and electron microscopy correlates structure of fluid catalytic cracking particles to Brønsted acidity. *Angewandte Chemie International Edition*, *51*(6), 1428–1431.

Karreman, M.A., Agronskaia, A.V., van Donselaar, E.G., Vocking, K, Fereidouni, F, Humbel, B.M., et al. (2012). Optimizing immuno-labeling for correlative fluorescence and electron microscopy on a single specimen. *Journal of Structural Biology*, *180*(2), 382–386.

Karreman, M.A., Van Donselaar, E.G., Agronskaia, A.V., Verrips, C.T., & Gerritsen, H.C. (2013). Novel contrasting and labeling procedures for correlative microscopy of thawed cryosections. *Journal of Histochemistry & Cytochemistry*, *61*(3): 236–47.

Microscopy Valley Project: http://www.stw-microscopyvalley.nl.

# Cellular Membrane Rearrangements Induced by Hepatitis C Virus

**Inés Romero-Brey\*, Carola Berger, Stephanie Kallis, Volker Lohmann, Ralf Bartenschlager**
Department of Infectious Diseases, Molecular Virology, Heidelberg University, Im Neuenheimer Feld 345, 69120, Heidelberg, Germany
\*Corresponding author: e-mail address: ines_romero-brey@med.uni-heidelberg.de

All positive-strand RNA viruses replicate in the cytoplasm in distinct membranous compartments serving as replication factories. Membranes building up these factories are recruited from different sources and serve as platforms for the assembly of multi-subunit protein complexes (the *replicase*) that catalyze the amplification of the viral RNA genome. In this study, we found that hepatitis C virus (HCV), a major causative agent of chronic liver disease, induces profound remodeling of primarily endoplasmic reticulum (ER)–derived membranes. By using correlative light and electron microscopy (CLEM), we observed that HCV triggers the formation of double membrane vesicles (DMVs), surrounding lipid droplets and residing in close proximity of the ER (Figure 9A; Romero-Brey *et al.*, 2012). Furthermore, by means of electron tomography (ET), we showed that these DMVs emerge as protrusions from ER tubules (Figure 9B; Romero-Brey *et al.*, 2012). CLEM allowed us to confirm the important contribution of one of the HCV nonstructural proteins (NS5A) to the formation of DMVs (Figure 10A; unpublished data). Importantly, inhibitors that are currently tested in clinical trials and targeting NS5A disrupt biogenesis of these HCV-induced mini-organelles and completely block virus replication (Figure 10B; Berger *et al.*, 2014).

These results unravel the mode of action of highly potent HCV inhibitors and disclose unexpected similarities between membranous replication factories induced by HCV and the very distantly related picornaviruses and coronaviruses.

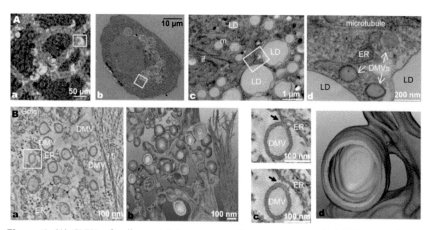

**Figure 9** (A) CLEM of cells containing a green fluorescent protein (GFP)–tagged HCV subgenomic replicon. (a) Epifluorescence microscopy of live cells containing a subgenomic replicon with a GFP-tagged NS5A, growing on sapphire discs with a carbon-coated coordinate pattern; (b) merge of EM and fluorescence images; (c) EM micrographs of the boxed cell region in panel (b), corresponding to an LD-enriched area containing DMVs in very close proximity to the ER. (b) ET of HCV-infected cells. (a) Slice of a dual axis tomogram showing the various membrane alterations and (b) 3D model of the entire tomogram; (c) serial single slices through the DMV boxed in panel (a) displaying a connection between the outer membrane of a DMV and the ER membrane (black arrows); (d) 3D surface model showing the membrane connection. LD, lipid droplet; ER, endoplasmic reticulum; DMV, double membrane vesicle; m, mitochondrium; if, intermediate filament. *Adapted from Romero-Brey* et al. *(2012).*

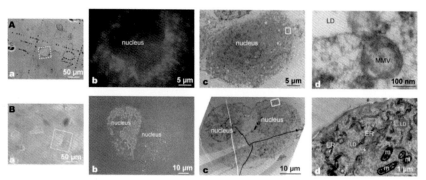

**Figure 10** (A) CLEM of cells expressing NS5A-RFP. (a) and (b) Light microscopy of cells expressing NS5A tagged with RFP; (c) and (d) electron micrographs of the cell highlighted with a dashed box in panel (a), depicting the formation of multimembrane vesicles resembling DMVs. (B) Effect of BMS-553 treatment (an anti-NS5A inhibitor) on DMV formation. (a) and (b) Light microscopy of cells expressing the NS3-NS5A HCV polyprotein tagged with GFP and pretreated (prior to transfection) with BMS-553; (c) and (d) electron micrographs of the cell highlighted with a dashed box in panel (a), showing that cells treated with this compound do not show any DMV.

# REFERENCES

Berger, C; Romero-Brey, I., Radujkovic, D., Terreux, R., Zayas, M., Paul, D., et al. (2014). Daclatasvir-like inhibitors of NS5A block early biogenesis of HCV-induced membranous replication factories, independent of RNA replication. *Gastroenterology*, July 18.

Romero-Brey, I., Merz, A., Chiramel, A., Lee, J.Y., Chlanda, P., Haselman, U., et al. (2012). Three-dimensional architecture and biogenesis of membrane structures associated with hepatitis C virus replication. *PLoS Pathogens*, *8,* e1003056.

SESSION 2

# In Situ Observations of Biomineralization Processes

# Nucleation and Particle Mediated Growth in Mineral Systems Investigated by Liquid-Phase TEM

**J.J. De Yoreo[a,]\*, M.H. Nielsen[b,c], Dongsheng Li[a], P.J.M. Smeets[c,d], N.A.J.M. Sommerdijk[c]**

[a]Physical Sciences Division, Pacific Northwest National Laboratory, Richland, WA 99352
[b]Department of Materials Science and Engineering, University of California, Berkeley, 94720
[c]Molecular Foundry, Lawrence Berkeley National Laboratory, Berkeley, CA 94720
[d]Laboratory of Materials and Interface Chemistry, Eindhoven University, Eindhoven, the Netherlands
\*Corresponding author: e-mail address: james.deyoreo@pnnl.gov

Solution-based growth of single crystals through assembly of nanoparticle precursors is a pervasive mechanism in many materials and mineral systems. Morever, the dominance of particle-mediated growth processes increases the importance of understanding the mechanisms and controls on nucleation of the precursor particles. Yet many longstanding questions surrounding nucleation remain unanswered and the postnucleation assembly process is poorly understood, due in part to a lack of experimental tools that can probe the dynamics of synthetic processes in liquids with adequate spatial and temporal resolution. Here, we report the results of using fluid cell TEM to investigate nucleation of calcium carbonate (Nielsen et al., 2014a) and particle assembly in both the iron oxyhydroxide (Li et al., 2012) and calcium carbonate systems (Nielsen et al., 2014b).

To examine nucleation of calcium carbonate, we used a custom-built fluid holder that enabled us to mix two reagents near the entrance to the cell and thus explore a wide range of solution conditions[1]. We observed the formation of amorphous calcium carbonate (ACC) over the entire range of conditions. In addition, we found that all common crystalline phases of calcium carbonate, including calcite, vaterite, and aragonite could form directly. Multiple phases often formed within a single experiment and the direct formation of the crystalline phases occurred under conditions in which ACC also readily formed. These observations demonstrate that multiple phases of calcium carbonate can form directly from solution without the intermediate stage of ACC. For all phases measured, we found

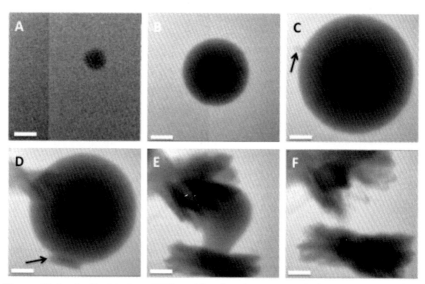

**Figure 11** In situ liquid phase TEM enables the observation of phase evolution during nucleation (Nielsen *et al.*, 2014a). (A-F) Time series showing CaCO₃ nucleation via a two-step process. The first phase to form is amorphous CaCO₃ (ACC) (A,B). This is followed by surface nucleation of aragonite (C, D) and consumption of the ACC (E, F). Just before the moment of aragonite nucleation, the ACC partcle shrinks in size, indicating expulsion of its structural water. Scale bar: 500 nm.

radial/edge growth rates after nucleation were linear with respect to time, showing that growth was reaction limited. Beyond these direct formation pathways, we observed transformation from ACC to aragonite and vaterite, but, significantly, not to calcite (Figure 11). In these observations, ACC transformed directly to the crystalline phases rather than undergoing a process of dissolution and reprecipitation. Nucleation of the second phase began on or just below the surface of the ACC particle and was preceded by a brief period of particle shrinkage, perhaps associated with expulsion of water. These formation pathways were confirmed by collecting diffraction information of the various phases of calcium carbonate.

To understand how the introduction of an organic matrix, which is common in biomineral systems, affects the nucleation of calcium carbonate, we performed a similar set of experiments in solutions containing the polyelectrolyte polystyrene sulfonate (PSS). Here, the cell was initially filled with CaCl₂ solution through one inlet and carbonate ions were introduced by diffusion from an ammonium carbonate source through the second inlet. In the absence of PSS, vaterite formed randomly throughout the fluid cell.

When PSS was introduced, it complexed more than half of the $Ca^{2+}$ ions and formed a globular phase. As carbonate diffused into the cell, the first solid phase to appear was ACC, which nucleated only within the globules. These results demonstrate that ion binding can play a significant role in directing nucleation, independent of any control over the free energy barrier to nucleation, which is usually inferred to be the primary factor leading to matrix-controlled nucleation (Habraken *et al.*, 2013).

We investigated the postnucleation growth of iron oxyhydroxide (Li *et al.*, 2012) and calcium carbonate (Nielsen *et al.*, 2014b) though particle assembly processes using a custom-built static fluid cell that enabled sub-nanometer resolution. We found that primary particles of ferrihydrite interacted with one another through translational and rotational diffusion until a near-perfect lattice match was obtained either with true crystallographic alignment or across a twin plane (Figure 12). Oriented attachment (OA) then occurred through a sudden jump-to-contact, demonstrating the existence of an attractive potential driving the OA process. Following OA, the resulting interfaces expanded through ion-by-ion attachment at a curvature-dependent rate. However, when a significant mismatch existed between the sizes of the two particles and attachment failed to occur over extended periods of particle interaction, the larger crystals still grew in size through Ostwald ripening, resulting in the disappearance of the smaller ones. In contrast to the clear role played by OA in the case of ferrihydrite, analysis of the assembly of akaganeite nanorods to form single-crystal hematite spindles showed that attachment did not result in coalignment; rather, the initial mesocrystal was disordered and recrystallized over time to become a well-ordered single crystal. Calcium carbonate exhibited still a different style of particle-mediated growth. In this system, we also observed that nanoparticles interacted and underwent aggregation events; however, the smallest particles often appeared to be amorphous, with crystallinity presumably arising as a result of attachment to the larger crystalline particle (Figure 13).

The results presented here highlight the wide array of pathways that are accessible during the nucleation process, as well as the diversity of mechanisms possible in particle mediated growth of single crystals. In both cases, the availability of liquid-phase TEM opens up new opportunities to decipher these underlying pathways and mechanisms.

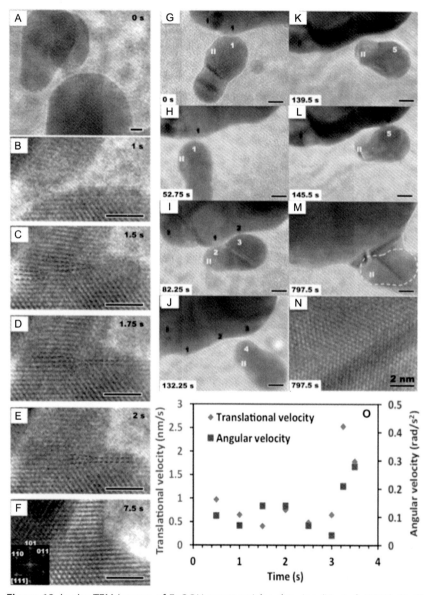

**Figure 12** In situ TEM images of FeOOH nanoparticles showing (Li *et al.*, 2012): (A–F) attachment at lattice-matched interface. Red dashed lines (C–E) highlight edge dislocation that translates to the right, leaving behind a defect-free interface (F). (G–M) Dynamics of attachment process. (N) Interface in (M) showing inclined (101) twin plane. Yellow dashed line (M) gives original boundary of attached particle. (O) Plot of relative translational and angular speeds leading up to attachment showing sudden acceleration over the last 5–10 Å.

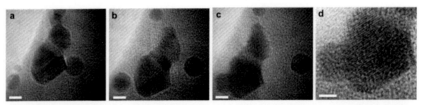

**Figure 13** (A–C) In situ TEM images showing 2–5 nm CaCO₃ amorphous particles fusing with larger crystalline mass (Nielsen *et al.,* 2014b). (D) Magnified region from (C) showing lattice fringes in crystalline particle. Scale bar: (A–C) 4 nm, (D) 2 nm, (E–H) 50 nm. Times are (A) 0, (B) 127.5 s, (C) 129.25 s.

## REFERENCES

Habraken, W.J.E.M., Tao, J., Brylka, L.J., Friedrich, H., Schenk, A.S., Verch, A., et al. (2013). Ion-association complexes unite classical and nonclassical theories for the biomimetic nucleation of calcium phosphate. *Nature Communications, 4,* 1507.

Li, D., Nielsen, M.H., Lee, J.R.I., Frandsen, C., Banfield, J.F., & De Yoreo, J.J. (2012). Direction-specific interactions control crystal growth by oriented attachment. *Science, 336,* 1014–1018.

Nielsen, M.H., Aloni, S., & De Yoreo, J.J. (2014a). In situ TEM imaging of CaCO₃ nucleation reveals coexistence of direct and indirect pathways. *Science, 345,* 1158–1162.

Nielsen, M.H., Li, D., Aloni, S., Han, T.Y.J., Frandsen, C., Seto, J., et al. (2014b). Investigating processes of nanocrystal formation and transformation via liquid cell TEM. *Microsccopy and Microanalysis,* **20,** 425–436.

# Studying the In Situ Growth and Biodegradation of Inorganic Nanoparticles by Liquid-Cell Aberration Corrected TEM

Damien Alloyeau[a,*], Yasir Javed[a], Walid Darchaoui[a], Guillaume Wang[a], Florence Gazeau[b], Christian Ricolleau[a]

[a]Laboratoire Matériaux et Phénomènes Quantiques, CNRS—Université Paris Diderot, France
[b]Laboratoire Matières et Systèmes Complexes, CNRS—Université Paris Diderot, France
*Corresponding author: e-mail address: damien.alloyeau@univ-paris-diderot.fr

Using liquid-cell TEM holder in an aberration-corrected TEM is a major technological rupture for understanding the complex phenomena arising at the liquid/solid interface. Recent microelectromechanical system (MEMS)-based technology allows imaging the dynamics of nano-objects in an encapsulated liquid solution within an electron-transparent microfabricated cell. The environmental conditions are finely controlled with a micro-fluidic system which enables to mix different reaction solutions at the observation window. Here, we performed the direct in situ study of two crucial phenomena in materials science: (1) the growth mechanisms of gold nanoparticles (NPs). (2) The degradation mechanisms of iron oxide NPs in a solution mimicking cellular environment.

We studied the growth of gold NPs *via* the reduction of metal salt. These growth mechanisms observed with a resolution below 0.2 nm, is indirectly induced by the electron beam. Indeed, 200-kV incident electrons radiolyze the water, creating free radicals and aqueous electrons that reduce metallic precursors. We have shown that the growth mode of gold NPs depends highly on the electron dose. High electron doses result in a diffusion–limited growth mode, leading to large dendritic structures, while low electron dose allows the formation of faceted NPs due to reaction-limited growth. These

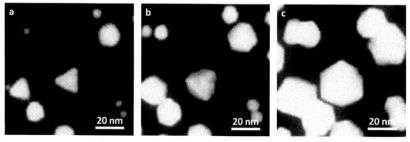

**Figure 14** In situ growth of facetted gold NPs observed by low-dose STEM-HAADF. We observe a shape transition between two nano-polyhedra.

latter conditions enable the fascinating study of the 2D growth mechanisms of nanoplates. (Figure 14).

If the understanding of the formation mechanisms of inorganic NPs is very important for controlling upstream their shape related–properties, studying their reactivity and transformation mechanisms in cellular environment is essential for evaluating their long–term efficiency as diagnostic or therapeutic agents. Here, we demonstrated for the first time that liquid–cell TEM is a relevant method to follow the (bio)degradation of iron oxide NPs in a solution mimicking the intracellular environment to which they are exposed during their life cycle in the organism (Figure 15).

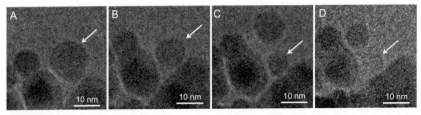

**Figure 15** In situ follow-up of the degradation of iron-oxide NPs. The corrosion and dissolution of a single NP (white arrow) is directly observed in a solution mimicking the intracellular environment. Observation time: (A) 0 s, (B) 220 s, (C) 540 s (an additional NP appeared in the field of view by diffusion), (D) 1080 s.

# In Situ TEM Shows Ion Binding Is Key to Directing CaCO$_3$ Nucleation in a Biomimetic Matrix

**P.J.M. Smeets[a,b,c,*], K.R.Cho[b], R.G.E. Kempen[a], N.A.J.M Sommerdijk[a], J.J. De Yoreo[c]**

[a]Eindhoven University of Technology, Eindhoven, Netherlands
[b]Lawrence Berkeley National Laboratory, Berkeley, CA
[c]Pacific Northwest National Laboratory, Richland, WA
*Corresponding author: e-mail address: P.J.M.Smeets@tue.nl

Biominerals possess shapes, structures, and properties not found in synthetic minerals. These defining characteristics arise from the interplay of the mineral with a macromolecular matrix, which directs crystal nucleation and growth (Lowenstam & Weiner, 1989; Mann, 2001). Within this three-dimensional biomolecular assembly, the developing mineral interacts with acidic macromolecules, either dissolved in the crystallization medium or associated with insoluble framework polymers such as chitin or collagen (Palmer *et al.*, 2008). Although acidic macromolecules are known to affect growth habits and phase selection, or even to completely inhibit precipitation in solution (Sommerdijk & With, 2008; Meldrum & Cölfen, 2008; Gower, 2008), little is known about the role of matrix-immobilized acidic macromolecules in directing mineralization.

This lack of understanding is, in part, due to the difficulty of studying biomimetic mineralization systems with sufficient spatial and temporal resolution (Dey *et al.*, 2010). However, liquid phase transmission electron microscopy (LP-TEM) can visualize events in situ in thin liquid volumes (about 500 nm in height) confined within two electron transparent silicon nitride (SiN) membranes. Here, we use LP-TEM to visualize the nucleation and growth of CaCO$_3$ in a biomimetic matrix of polystyrene sulfonate (PSS). In particular, we utilized a dual inlet flow stage where we started out with a CaCl$_2$ solution in the confined cell, after which carbonate was introduced through in-diffusion of vapor released from the decomposition of solid (NH$_4$)$_2$CO$_3$ (mainly CO$_{2\,(g)}$ and NH$_{3\,(g)}$) via the second inlet port. Within minutes, we directly observed the nucleation of randomly distributed vaterite nanoparticles over the surface, which we attribute to heterogeneous nucleation on the SiN.

When we introduced a PSS solution together with our CaCl$_2$ solution, in situ imaging of this model system showed that the PSS by binding of Ca$^{2+}$

ions is able to form globules of about 10–100 nm that can adsorb onto the SiN surface, as shown by our previous work (Smeets *et al.*, 2013). We find that calcium is able to bind to the sulfonate groups of the polymer (as determined by FTIR measurements); $56 \pm 6$ mole % of the total added calcium is bound as determined by performing $Ca^{2+}$ ion selective electrode measurements.

The immobilized Ca–PSS globules inside the TEM cell were subsequently exposed to vapor from the decomposition of solid $(NH_4)_2CO_3$. After a delay time of about 20 min the nucleation of $CaCO_3$ nanoparticles was observed, growing to sizes of 10–20 nm within seconds (Figure 16A,C). In situ electron diffraction showed that the as-formed particles were amorphous (Figure 16C, inset). Growth rate profiles of ACC (with PSS) and vaterite (without PSS) were extracted from the time-lapse series, where the initial rates were higher for ACC particles (16–23 nm/s) compared to vaterite crystallites (3–6 nm/s) (Figure 16B). To validate that the observed nucleation of nanoparticles was not the result of the continuous exposure to

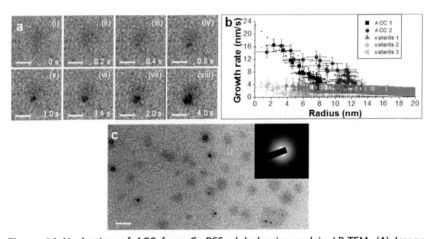

**Figure 16** Nucleation of ACC from Ca-PSS globules imaged in LP-TEM. (A) Image sequence after 45 min of $(NH_4)_2CO_3$ diffusion showing initial nucleation and growth of a $CaCO_3$ particle inside (i) a primary Ca-PSS globule within 4 s (viii) (ACC 1, scale bar: 20 nm). (B) Extrapolated growth rates versus average radius for two ACC particles (ACC 1 & 2 with PSS) compared to those of three vaterite particles (without PSS). A logarithmic function is used to fit each set of data points and to extrapolate to zero radius. (C) Lower magnification (scale bar: 50 nm) image exhibiting many nuclei in Ca-PSS globules and an electron diffraction pattern (inset) showing they are amorphous. Differences in the radial distribution of black and white contrast in the nucleated ACC particles is due to differences in the location relative to the focal plane: Black signifies near-focus condition (near the bottom of the SiN membrane) and white reflects over-focus (near the top of the membrane) (Smeets *et al.*, 2015).

the electron beam, the reaction solution was imaged at different time points between 20 and 65 min at unexposed areas using a single exposure with an electron dose of about 50–300 e/$\mathring{A}^2$, comparable to what is used in low dose cryo-TEM imaging (Friedrich *et al.*, 2010). In these images, comparable amorphous nanoparticles were identified, again forming only at the sites of Ca-PSS globules.

Thus, we verify that the binding of $Ca^{2+}$ in the globules is a key step in the controlled formation of metastable amorphous calcium carbonate, an important precursor phase in many biomineralization systems (Addadi *et al.*, 2003). These findings now reveal the significant role that ion binding can play in directing nucleation independent of any controls over the free energy barriers.

## REFERENCES

Addadi, L., Raz, S., & Weiner, S. (2003). Taking advantage of disorder: Amorphous calcium carbonate and its roles in biomineralization. *Advanced Materials*, *15*, 959–970.

Dey, A., de With, G., & Sommerdijk, N.A.J.M. (2010). In situ techniques in biomimetic mineralization studies of calcium carbonate. *Chemical Society Reviews*, *39*, 397–409.

Friedrich, H., Frederik, P.M., de With, G., & Sommerdijk, N.A.J.M. (2010). Imaging of self-assembled structures: Interpretation of TEM and cryo-TEM images. *Angewandte Chemie International Edition*, *49*, 7850–7858.

Gower, L.B. (2008). Biomimetic model systems for investigating the amorphous precursor pathway and its role in biomineralization. *Chemical Reviews*, *108*, 4551–4627.

Lowenstam, H.A., & Weiner, S. (1989). *On Biomineralization*. Oxford University Press, Oxford, UK.

Mann, S. (2001). *Biomineralization, Principles and Concepts in Bioinorganic Materials Chemistry*. Oxford University Press, Oxford, UK.

Meldrum, F.C., & Cölfen, H. (2008). Controlling mineral morphologies and structures in biological and synthetic systems. *Chemical Reviews*, *108*, 4332–4432.

Palmer, L.C., Newcomb, C.J., Kaltz, S.R., Spoerke, E.D., & Stupp, S.I. (2008). Biomimetic systems for hydroxyapatite mineralization inspired by bone and enamel. Special issue on biomineralization. *Chemical Reviews*, *108*, 4329–4978.

Smeets, P.J.M., Cho, K.R., Kempen, R.G.E., Sommerdijk, N.A.J.M., & De Yoreo, J.J. (2015). Calcium carbonate nucleation driven by ion binding in a biomimetic matrix revealed by in situ microscopy. *Nature Materials*, early online, http://dx.doi.org/10.1038/nmat4193.

Smeets, P.J.M., Li, D., Nielsen, M.H., Cho, K.R., Sommerdijk, N.A.J.M., & De Yoreo, J.J. (2013). Unraveling the $CaCO_3$ mesocrystal formation mechanism including a polyelectrolyte additive using in situ TEM and in situ AFM. *Advances in Imaging and Electron Physics*, *179*, 165–167.

Sommerdijk, N., & de With, G. (2008). Biomimetic $CaCO_3$ mineralization using designer molecules and interfaces. *Chemical Reviews*, *108*, 4499–4550.

# Crystallisation of Calcium Carbonate Studied by Liquid Cell Scanning Transmission Electron Microscopy

**Andreas Verch[a,b,*], Iva Perovic[c], Ashit Rao[d], Eric P. Chang[a], Helmut Cölfen[d],
John Spencer Evans[c], Roland Kröger[b]**
[a]Department of Physics, University of York, Heslington, York, YO10 5DD, UK
[b]INM-Leibniz Institute for New Materials, Campus D2 2, 66123 Saarbrücken, Germany
[c]Laboratory for Chemical Physics, Division of Basic Sciences and Center for Skeletal Biology, New York
University, 345 E. 24th Street, NY, NY, 10010
[d]Department of Chemistry, Physical Chemistry, University Konstanz, Universitätsstraße 10, 78457 Konstanz,
Germany
*Corresponding author: e-mail address: andreas.verch@inm-gmbh.de

Biological systems and many industrial applications depend on a precise control of crystallisation processes. In living systems, a variety of proteins regulate the progress of the mineralisation leading to materials with enhanced properties (Arias & Fernández, 2008). In order to improve biomimetic materials engineering approaches, it is necessary to gain a better understanding of how organics interact with inorganics in the process of particle formation.

In this work, we focus on the impact of the protein AP7 on the in vitro crystallisation of calcium carbonate. AP7 is an intracrystalline protein found in the nacre layer of the pacific red abalone (Michenfelder *et al.*, 2003). It forms intrinsically instable oligomers that assemble into larger aggregates when calcium ions are present (Amos *et al.*, 2011). Our potentiometric experiments show that this protein influences the progress of the crystallization already in the prenucleation stage by associating with calcium carbonate prenucleation species, thereby prolonging the time till the onset of nucleation (Perovic *et al.*, 2014). Protein phase–calcium carbonate nanoparticle interactions in the postnucleation stage were observed by a conventional TEM time series study showing entrapped calcium carbonate particles within the protein phase. However, conventional TEM requires a high vacuum that commonly leads to a number of undesired drying artefacts.

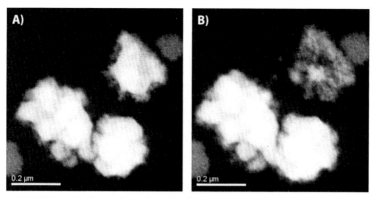

**Figure 17** (A) Crystals formed in a liquid cell STEM experiment in the presence of AP7. The fringed structures at the crystal edges are formed of AP7. (B) The crystalline particle in the top-right corner dissolved in favor of the other two crystals with only the organic protein framework remaining.

Fluid cell scanning transmission electron microscopy allowed us to overcome this shortcoming and provided additional time-resolved information of the particle formation process. For these experiments, we utilized a commercially available three-port liquid cell holder where the liquid is situated between two silicon chips with electron transparent silicon nitride windows in the centre. Solutions of calcium ions and bicarbonate/carbonate ions, respectively, were mixed within the TEM holder tip in order to see the precipitation reaction from the onset. Using the electron microscope in scanning mode enabled us to image through liquid thicknesses of a few microns and moreover restrict the area exposed to the electron beam only to the investigated location. In this study, we found that AP7 assembles and forms calcium rich-nanocrystal-protein networks if both calcium and carbonate ions were present. These protein networks are incorporated into the growing calcium carbonate crystals and even remain unchanged when the surrounding crystal is dissolved by a variation of the experimental conditions (Figure 17).

## ACKNOWLEDGMENTS

H.C. and A.R. thank the Konstanz Research School Chemical Biology for a PhD stipend to A.R. We acknowledge the York JEOL Nanocentre for the electron microscope facilities and support for A.V. and R.K. by the Engineering and Physical Sciences Research Council [grant number EP/I001514/1]. Research related to NMR, SEM, TEM, mineralization assays, and AFM experiments were supported by the U.S. Department of Energy, Office of Basic Energy Sciences, Division of Materials Sciences and Engineering under Award DE-FG02-03ER46099.

# REFERENCES

Amos, F.F., Ndao, M., Ponce, C.B., & Evans, J.S. (2011). A C-RING-like domain partic-
ipates in protein self-assembly and mineral nucleation. *Biochemistry, 50,* 8880–8887.
Arias, J.L., & Fernández, M.S. (2008). Polysaccharides and proteoglycans in calcium
carbonate-based biomineralization. *Chemical Reviews, 108,* 4475–4482.
Michenfelder, M., Fu, G., Lawrence, C., Weaver, J.C., Wustman, B.A., Taranto, L., et al.
(2003). Characterization of two molluscan crystal-modulating biomineralization proteins and identification of putative mineral binding domains. *Biopolymers, 70,* 522–533.
Perovic, I., Verch, A., Chang, E.P., Rao, A., Cölfen, H., Kröger, R., & Evans, J.S. (2014).
An oligomeric C-RING nacre protein influences prenucleation events and organizes
mineral nanoparticles. *Biochemistry, 53,* 7259–7268.

# Designing In Situ Experiments

# Studies of Transport Properties Using In Situ Microscopy

**Eva Olsson***

Department of Applied Physics, Chalmers University of Technology, 412 96 Gothenburg, Sweden
*Corresponding author: e-mail address: eva.olsson@chalmers.se

As material synthesis and nanofabrication methods are refined and the control of material structure reaches beyond the nanoscale, the role of individual interfaces, defects, and atoms is pronounced and can dominate the properties. The strong influence of local atomic structure offers the possibility of designing new components with tailored and unique properties. Electron microscopes offer the possibility of correlating the structure to transport properties with a spatial resolution that reaches the atomic scale. A knowledge platform of how to design new materials by combining experiments and theory can thus be established. In addition, inserting a scanning tunnelling or an atomic force microscope in the electron microscope enables studies of dynamic events. This talk will cover aspects of in situ studies using manipulators with high spatial precision in electron microscopes, including experiments in low-vacuum conditions and give examples from different material systems.

# Calibrated In Situ Transmission Electron Microscopy for the Study of Nanoscale Processes in Liquids

Patricia Abellan[a,]*, Taylor J. Woehl[b], James E. Evans[c], Nigel D. Browning[a]

[a]Fundamental and Computational Sciences Directorate, Pacific Northwest National Laboratory, P.O. Box 999, Richland, WA 99352
[b]Division of Materials Science and Engineering, U.S. DOE Ames Laboratory, Ames, IA 50011
[c]Environmental Molecular Sciences Laboratory, Pacific Northwest National Laboratory, P.O. Box 999, Richland, WA 99352
*Corresponding author: e-mail address: pabellan@superstem.org

Using fluid stages in the TEM, in situ observations of dynamic processes in liquids can be made with nanometer and even atomic spatial resolutions. The flexibility of this approach for studying liquid phase reactions relies upon both the possibility of obtaining information by using well established as well as novel capabilities in the TEM platform and the possibility of fine-tuning the design of the experiment. Fluid stages are designed to fit in any TEM, allowing for local chemical information in microscopes equipped with electron energy loss spectrometers, sub-angstrom spatial resolution using Cs-correction, or high temporal resolution on the 10 ns to 1 μs range, if using a dynamic TEM (DTEM). Additional flexibility comes from the design of the fluid stage, where the liquid chamber is created and separated from the high vacuum in the microscope column using two microfabricated chips. Thus, the experiment layout can be adapted for each sample by using chips designed for a specific application including different configurations of chips for continuous flow applications, mixing of two chemicals or with electrical contacts. As the technique is increasingly refined, the challenge now becomes obtaining quantifiable and reproducible data that is free from beam-induced artifacts. Once the effect of the electron beam on the liquid sample is calibrated, reliable information on the fundamental mechanisms behind the colloidal synthesis of nanostructured materials, self-assembly or the structural processes during battery operation, and even biological conformational changes can be uncovered, just to name a few.

Studying liquid samples in the TEM represents specific challenges as compared to the study of crystalline/amorphous solid specimens. Solutions decompose upon radiolysis and chemical species are generated in the liquid phase and interact with the sample, most times beyond the observation area since they are mobile by nature. Factors such as the electron dose applied to the sample (Zheng *et al.*, 2009; Evans *et al.*, 2011; Woehl *et al.*, 2012; Grogan *et al.*, 2014), accelerating voltage (Abellan *et al.*, 2014a) and imaging mode (e.g., TEM, STEM, SEM; Abellan *et al.*, 2014a; de Jonge & Ross, 2011), liquid thickness (Jungjohann *et al.*, 2013), and solution composition (Evans *et al.*, 2011; Schneider *et al.*, 2014) strongly determine the amount of radiation damage produced, and their effect must be calibrated for. In many systems, as for the case of an aqueous Ag precursor solution, radiolysis of deionized water in the TEM leads to the nucleation and growth of particles (Woehl *et al.*, 2012; Abellan *et al.*, 2014a). As an example, Figure 18 illustrates the effect of increasing electron dose and kilovolts on the colloidal growth of silver in a fluid stage. The seven different bright field (BF) STEM images are snapshots of in situ movies recorded for increasing electron dose values (left to right) and increasing electron beam energy (top versus bottom row) after the same irradiation time [except for Figure 18 (D)]. The different growth regimes observed at high energy–reaction and diffusion-limited growth as the electron dose values increase (Woehl *et al.*, 2012) could not be reproduced at 80 kV, suggesting a stronger overall effect of oxidizing radicals for lower energies (Abellan *et al.*, 2014a). For the specific case of growth of nanomaterials, using solvents with higher chemical complexity, such as organic solvents, leads to the production of a larger variety of radicals

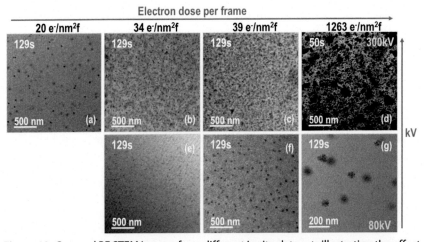

**Figure 18** Cropped BF STEM images from different in situ data sets illustrating the effect of electron dose and beam energy on the electron-beam induced growth of Ag nanocrystals from solution.

(as compared to water) increasing the number of chemical reactions involved. Therefore, understanding how these solutions are affected by the electron beam becomes even more crucial. Besides the study of particle growth, the damaging effect of the electron beam can be used for other applications, such as an effective tool for studying the degradation mechanisms of electrolyte solutions used for Li-ion battery technology (Abellan et al., 2014b). This approach provides the possibility of rapidly screening electrolytes candidates for an improved stability upon reduction. For all of this, we believe that as the interest on understanding the interaction of the electron beam with a higher number of liquid systems increases so too do the possibilities of application and the range of dynamic processes that can be explored.

## ACKNOWLEDGMENTS

This work was supported by the Chemical Imaging Initiative; under the Laboratory Directed Research and Development Program at Pacific Northwest National Laboratory (PNNL). PNNL is a multiprogram national laboratory operated by Battelle for the U.S. Department of Energy (DOE) under Contract DE-AC05-76RL01830. A portion of the research was performed using the Environmental Molecular Sciences Laboratory (EMSL), a national scientific user facility sponsored by the Department of Energy's Office of Biological and Environmental Research and located at Pacific Northwest National Laboratory.

## REFERENCES

Abellan, P., Woehl, T.J., Parent, L.R., Browning, N.D., Evans, J.E., & Arslan, I. (2014). Factors influencing quantitative liquid (scanning) transmission electron microscopy. *Chemical Communications, 50*(38), 4873-4880.

Abellan, P., Mehdi, B.L., Parent, L.R., Gu, M., Park, C., Xu, W., et al. (2014). Probing the degradation mechanisms in electrolyte solutions for Li-ion batteries by in situ transmission electron microscopy. *Nano Letters, 14*(3), 1293–1299.

Evans, J.E., Jungjohann, K.L., Browning, N.D., & Arslan, I. (2011). Controlled growth of nanoparticles from solution with in situ liquid transmission electron microscopy. *Nano Letters, 11*(7), 2809–2813.

Grogan, J.M., Schneider, N.M., Ross, F.M., & Bau, H.H. (2014). Bubble and pattern formation in liquid induced by an electron beam. *Nano Letters, 14*(1), 359–364.

Jungjohann, K.L., Bliznakov, S., Sutter, P.W., Stach, E.A., & Sutter, E.A. (2013). In situ liquid cell electron microscopy of the solution growth of Au–Pd core–shell nanostructures. *Nano Letters, 13*(6), 2964–2970.

de Jonge, N., & Ross, F.M. (2011). Electron microscopy of specimens in liquid. *Nature Nanotechnology, 6*(11), 695–704.

Schneider, N.M., Norton, M.M., Mendel, B.J., Grogan, J.M., Ross, F.M., & Bau, H.H. (2014). Electron–water interactions and implications for liquid cell electron microscopy. *Journal of Physical Chemistry C, 118*(38), 22373–22382.

Woehl, T.J., Evans, J.E., Arslan, I., Ristenpart, W.D., & Browning, N.D. (2012). Direct in situ determination of the mechanisms controlling nanoparticle nucleation and growth. *ACS Nano, 6*(10), 8599–8610.

Zheng, H.M., Claridge, S.A., Minor, A.M., Alivisatos, A.P., & Dahmen, U. (2009). Nanocrystal diffusion in a liquid thin film observed by in situ transmission electron microscopy. *Nano Letters, 9*(6), 2460–2465.

# Microchip Systems for In Situ Electron Microscopy of Processes in Gases and Liquids

Sardar Bilal Alam[a], Eric Jensen[a], Frances M Ross[b], Ole Hansen[a,d], Andy Burrows[c], Kristian Mølhave[a,*]

[a]DTU-Nanotech, Department of Micro and Nanotechnology, Technical University of Denmark (DTU), Denmark
[b]IBM Research Division, T.J. Watson Research Center, Yorktown Heights, NY
[c]DTU CEN, Center for Electron Nanoscopy, DTU
[d]CINF-Center for Individual Nanoparticle Functionality, DTU
*Corresponding author: e-mail address: Kristian.molhave@nanotech.dtu.dk

Techniques allowing real-time, high-resolution imaging of nanoscale processes by in situ TEM are currently under rapid development based on microchip systems that enable real-time imaging of controlled processes to be combined with, for instance, correlated electrical measurements (Huang et al., 2010; Kallesøe et al., 2012). In addition to gas-phase ETEM studies, microchip-based systems also allow the imaging of processes in liquids (Williamson et al., 2003) and recently with good resolution (Li et al., 2012; Liao et al., 2012). Here, we report on our recent advances in joule heated microcantilevers (Molhave et al., 2008) used for gas-phase in situ TEM studies of nanowire growth (Kallesøe et al., 2010) and devices, and on our efforts in creating a suspended microchannel system for TEM imaging of liquid samples and processes.

We created and characterized nanowire devices by growing silicon nanowires in situ TEM using the VLS mechanism (Ross, 2010), from one joule heated microcantilever to an adjacent cantilever, thereby forming a bridge. Building on our previous experiments, we studied the details of contact formation when a hot nanowire grows into contact with a similarly heated adjacent cantilever. A silicon-to-silicon junction forms as the catalytic gold diffuses away. The suspended microcantilevers system opens up novel ways to create electrically contacted nanowire devices (Kallesøe et al., 2012) in addition to allowing in situ TEM studies of the many simultaneous processes taking place upon contact formation and in situ electrical measurements carried out after connected nanowires were formed.

To study liquids encapsulated between membrane windows on microchips, the reported microchip devices today all involve two bonded chips with thin membranes of silicon dioxide or nitride (Jonge & Ross, 2011). To allow better control of the encapsulated volume and geometry, we

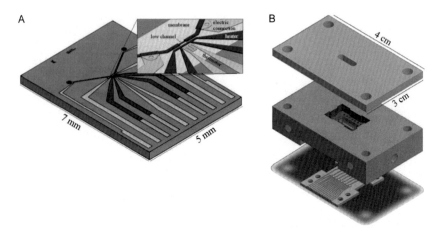

**Figure 19** (A) The TEM chip with 10 electrical connections. On the center of the chip is a 50-nm-thick suspended membrane with a microchannel and optional components such as heaters. (B) The ECSEM setup with the lid at the top, the chip holder in the middle, and electrical connection at the bottom.

are exploring devices using monolithic chips (Figure 19) with suspended microfluidic channels made from silicon nitride (Jensen et al., 2014). On the channels, thinned window regions allow higher resolution than in the supporting part of the channel. The system's novel geometry is expected to improve the imaging resolution by providing precise control over the channel height and the possibility of achieving small regions with ultrathin membranes for the best possible resolution. We have also developed an electrochemical setup suitable for use in SEM (Fig 2) (Jensen, Køblera, Jensen, & Mølhave, 2013).

## ACKNOWLEDGMENTS

The authors acknowledge the assistance of M.C. Reuter and A.W. Ellis of IBM and funding from FTP Case No. 10-082797 Nanolive, DFF- Sapere Aude LiquidEM, the Danish National Research Foundation's Center for Individual Nanoparticle Functionality (DNRF54), and DTU CEN.

## REFERENCES

Huang, J.Y., et al., (2010). In situ observation of the electrochemical lithiation of a single SnO2 nanowire electrode. *Science, 330*(6010), 1515–1520.

Jensen, E., Køblera, C., Jensen, P.S., & Mølhave, K. (2013). In situ SEM microchip setup for electrochemical experiments with water-based solutions. *Ultramicroscopy, 129,* 63–69.

Jensen, E., Burrows, A., & Mølhave, K. (2014). Monolithic chip system with a microfluidic channel for in situ electron microscopy of liquids. *Microscopy and Microanalysis, 20*(2), 445–451.

Jonge, N. de, & Ross, F.M. (2011). Electron microscopy of specimens in liquid. *Nature Nanotechnology*, *6*(11), 695–704.

Kallesøe, C., et al. (2012). In situ TEM Creation and electrical characterization of nanowire devices. *Nano Letters*, *12*(6), 2965–2970.

Kallesøe, C., Wen, C.Y., Mølhave, K., Bøggild, P., & Ross, F.M. (2010). Measurement of local Si-nanowire growth kinetics using in situ transmission electron microscopy of heated cantilevers. *Small*, *6*(18), 2058–2064.

Liao, H.-G., Cui, L., Whitelam, S., & Haimei Zheng, H. (2012). Real-time imaging of Pt3Fe nanorod growth in solution. *Science*, *336*(6084), 1011–1014.

Li, D., Nielsen, M.H., Lee, J.R. I., Frandsen, C., Banfield, J.F., & De Yoreo, J.J. (2012). Direction-specific interactions control crystal growth by oriented attachment. *Science*, *336*(6084), 1014–1018.

Molhave, K., Wacaser, B.A., Petersen, D.H., Wagner J.B., Samuelson L., & Bøggild, P. (2008). Epitaxial Integration of nanowires in microsystems by local micrometer-scale vapor-phase epitaxy. *Small*, *4*(10), 1741–1746.

Ross, F.M. (2010). Controlling nanowire structures through real time growth studies. *Reports on Progress in Physics*, *73*(11), p.114501.

Williamson, M.J., Tromp, R.M., Vereecken, P.M., Hull, R., & Ross, F.M. (2003). Dynamic microscopy of nanoscale cluster growth at the solid-liquid interface. *Nature Materials*, *2*(8), 532–536.

# Scanning Transmission Electron Microscopy of Liquid Specimens

**N. de Jonge[a,b,]\*, M. Pfaff[a], D.B. Peckys[a]**
[a]INM-Leibniz Institute for New Materials, Campus D2 2, 66123 Saarbrücken, Germany
[b]Department of Physics, University of Saarland, Campus A5 1, 66123 Saarbrücken, Germany
*Corresponding author: e-mail address: niels.dejonge@inm-gmbh.de

Traditionally, electron microscopy is used to study solid samples maintained in a vacuum. For a broad range of experiments, it is required to image samples in a liquid environment; for instance, for the study of nanoparticle growth, self-assembly processes, or catalytic nanoparticles, and for research on biological cells and macromolecules (de Jonge & Ross, 2011). A new option to study liquid samples of practical thicknesses of several micrometers was introduced in recent years, combining scanning transmission electron microscopy (STEM) with the usage of silicon nitride (SiN) membranes as windows of a microfluidic chamber named *Liquid STEM* (de Jonge *et al.*, 2009) (see Figure 20). Nanoscale resolution is achievable for nanomaterials of a high atomic number (Z) in a low–Z liquid resulting from the Z contrast of STEM. It is also possible to study thin liquid samples using the environmental scanning electron microscope (ESEM) with STEM detector.

Time-resolved Liquid STEM is feasible and can be used to study the movements of nanoparticles in liquid. We have found that the movement of nanoparticles in close proximity to the SiN membrane is up to three orders of magnitude slower than of Brownian motion in a bulk liquid (Ring & de Jonge, 2012). Liquid STEM is useful to explore growth processes of nanomaterials, such as gold dendrites (Kraus & de Jonge, 2013).

The capability to achieve nanoscale resolution in liquid is of great interest for the study of membrane proteins in cells, which is challenging because the involved nanoscale dimensions require a spatial resolution beyond that of state-of-the-art fluorescence microscopy, while cells have to be prepared into thin solid samples for conventional electron microscopy preventing the study of whole cells. An example involves the dimerization of the EGFR, a transmembrane receptor playing a critical role in the pathogenesis and progression of many different types of cancer. An important question is under which conditions and in which cellular regions dimerization occurs. Liquid STEM provided a spatial resolution of a few nanometers to identify the protein complex subunits in the images while the cell remains intact and

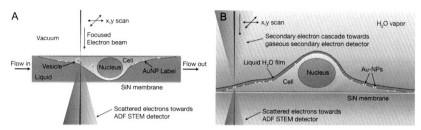

**Figure 20** Principles of Liquid STEM. A whole cell containing proteins labeled with gold nanoparticles (AuNPs) is imaged using the annular dark field (ADF) detector beneath the sample. (A) The cell is fully enclosed in a microfluidic chamber with two SiN windows for STEM. (B) The cell is maintained in a saturated water vapor atmosphere, while a thin layer of water covers the cell for ESEM-STEM. *Used with permission from Peckys & de Jonge (2014).* (See the color plate.)

in its natural aqueous environment (de Jonge *et al.*, 2009; Peckys *et al.*, 2013; Peckys & de Jonge, 2014). Correlative fluorescence microscopy and Liquid STEM of whole cells is readily possible via the usage of fluorescent nanoparticles as specific protein labels (Figure 21).

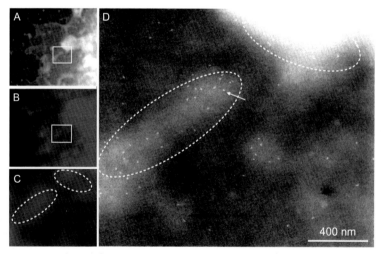

**Figure 21** Correlative fluorescence microscopy and ESEM-STEM of quantum dot labeled ErbB2 receptors on a SKBR3 cell. (A) Overview of ESEM-STEM image. (B) Fluorescence image recorded from the same area as shown in (A). (C) Enlarged detail from the image shown in (C) of the same size as (D). (D) STEM image from the boxed region shown in (A). The locations of individual HER2 receptors are indicated by the bright, bullet-shaped QDs (example at arrow). HER2 concentrates on electron denser (bright) areas (enclosed in dashed lines); i.e., membrane ruffles. (See the color plate.)

## ACKNOWLEDGMENTS

We thank M. Koch for help with the experiments, and Protochips Inc., for providing the microchips and liquid specimen holder. We thank E. Arzt for his support through INM. This research was supported in part by the Leibniz Competition 2014.

## REFERENCES

de Jonge, N., Peckys, D.B., Kremers, G.J., & Piston, D.W. (2009). Electron microscopy of whole cells in liquid with nanometer resolution. *Proceedings of the National Academy of Sciiences, 106,* 2159–2164.

de Jonge, N., & Ross, F.M. (2011). Electron microscopy of specimens in liquid. *Nature Nanotechnology, 6,* 695–704.

Kraus, T., & de Jonge, N. (2013). Dendritic gold nanowire growth observed in liquid with transmission electron microscopy. *Langmuir: The ACS Journal of Surfaces and Colloids, 29,* 8427–8432.

Peckys, D.B., Baudoin, J.P., Eder, M., Werner, U., & de Jonge, N. (2013). Epidermal growth factor receptor subunit locations determined in hydrated cells with environmental scanning electron microscopy. *Scientific Reports, 3,* 2626.

Peckys, D.B., & de Jonge, N. (2014). Liquid scanning transmission electron microscopy: Imaging protein complexes in their native environment in whole eukaryotic cells. *Microscopy and Microanalysis, 20,* 189–198.

Ring, E.A., & de Jonge, N. (2012). Video-frequency scanning transmission electron microscopy of moving gold nanoparticles in liquid. *Micron, 43,* 1078–1084.

# Scanning Electron Spectro-Microscopy in Liquids and Dense Gaseous Environment Through Electron Transparent Graphene Membranes

**Andrei Kolmakov***

Center for Nanoscale Science and Technology, NIST Gaithersburg, MD 20899–6204
*Corresponding author: e-mail address: andrei.kolmakov@nist.gov

Electron microscopy and spectroscopy of complex liquid–gas, solid–liquid, and solid–gas interfaces in situ under realistic conditions is largely demanded in bio-medical, energy, catalysis research, microelectronics, and environmental studies. The performance of the traditional environmental scanning electron microscopy (ESEM), as well as novel membrane-based environmental cells for atmospheric pressure SEM, depends on electron mean free path in the ambient media and/or membrane itself, which usually is very small in solids, liquids, and dense gases. Novel 2D materials such as graphene and its derivatives have only a single atomic layer or a few thin layers and possess the unique combination of the mechanical strength (Lee et al., 2008), molecular impermeability (Bunch et al., 2008), optical and electron transparency (Stoll et al., 2012) in a wide electron energy range (see Figure 22). The latter makes these materials as prospective electron windows for ambient pressure SEM and electron spectroscopy, as well as for correlative imaging. In this report, we describe few tested designs of the graphene based environmental cells (GE-cell) along with protocols for their high yield fabrication (Krueger et al., 2011). We compare these graphene windows with standard SIN membranes and demonstrate the capability to perform low-energy scanning electron microscopy (Stoll et al., 2012), Auger electron spectroscopy (AES) and X-ray photoelectron spectroscopy (XPS) (Kolmakov et al., 2011) at atmospheric pressure using GE cells, which are hard to accomplish using standard approaches. The process of water radiolysis and bubble formation at the water–graphene interface was studied in vivo using SEM and XPS to demonstrate the capabilities of the technique (Figure 23; Kraus et al., 2014).

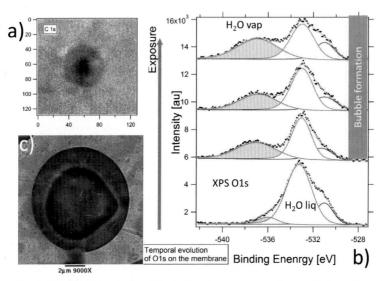

**Figure 22** (left) TPP_2M calculated inelastic electron mean free path in carbon. The thickness of one and two layers of graphene is depicted in the bottom of the graph. Inset: E-cell principle design of graphene E-cell and electron attenuation formula; (right) the comparative SEM image of a water-filled GE cell, with and without Au nanoparticles. (See the color plate.)

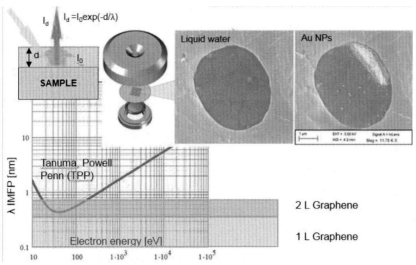

**Figure 23** (A) C1s XPS chemical map of the graphene-covered, water-filled micro-channel after local spectra acquisition; (B) Time evolution of the O1s spectra taken from the center of the membrane in (A); (C) SEM images of the bubble (darker area with lower electron yield) formed during e-beam irradiation of the water behind graphene membrane (see Stoll et al., 2012 for details).

# REFERENCES

Bunch, J.S., et al. (2008). Impermeable atomic membranes from graphene sheets. *Nano Letters, 8*(8), 2458–2462.

Kolmakov, A., et al. (2011). Graphene oxide windows for in situ environmental cell photoelectron spectroscopy. *Nature Nanotechnology, 6,* 651–657.

Kraus, J., et al. (2014). Photoelectron spectroscopy of wet and gaseous samples through graphene membranes. *Nanoscale, 6,* 14394–14403.

Krueger, M., et al. (2011). Drop-casted self-assembling graphene oxide membranes for scanning electron microscopy on wet and dense gaseous samples. *Acs Nano, 5*(12), 10047–10054.

*Lee, C., Wei, X., Kysar, J.W., & Hone, J. (2008). Measurement of the elastic properties and intrinsic strength of monolayer graphene. *Science, 321*(5887), 385–388.

Stoll, J.D., & Kolmakov, A. (2012). Electron transparent graphene windows for environmental scanning electron microscopy in liquids and dense gases. *Nanotechnology, 23*(50), 505704.

# High-Temperature and Other In Situ Experiments

# In situ HT-ESEM Observation of CeO$_2$ Nanospheres Sintering: From Neck Elaboration to Microstructure Design

**G.I. Nkou Bouala[a,*], R. Podor[a], N. Clavier[a], J. Lechelle[b], N. Dacheux[a]**

[a]ICSM, UMR 5257 CEA/CNRS/UM2/ENSCM, Site de Marcoule—Bât. 426, BP 17171, 30207 Bagnols/Cèze cedex, France
[b]CEA/DEN/DEC/SPUA/LMP, Site de Cadarache—Bât. 717, 13108 St-Paul lez Durance, France
*Corresponding author: e-mail address: GalyIngrid.nkoubouala@cea.fr

*Sintering* could be defined as the transformation of a powdered compact to a cohesive material under heating at high temperatures. It appears as a key-step in the preparation of ceramic materials such as nuclear fuels UOx and MOx (mixed oxide U and Pu). The sintering is usually described as having three different stages. The initial stage involves the elaboration of necks between the grains and leading to the cohesion of material, the intermediate and final stages are dedicated to the elimination of porosity between the grains by the means of grain growth mechanisms (Bernache-Assolant, 1993). The first stage of sintering is generally described through numerical simulation using simplified systems constituted by two or three spherical grains in contact (Wakai, 2006). Presently, only few experimental works are devoted to the kinetics of necks elaboration (corresponding to the first stage of sintering) using ex situ TEM (Qin *et al.*, 2010) and SEM (Slamovich & Lange, 1990) observations. These studies allowed experimental investigations of this stage of sintering on metallic materials with spherical grains.

In this study, we report the first experimental observations of the initial stage of sintering from grains of lanthanide oxides with controlled shapes by using environmental scanning electron microscopy (ESEM) at high temperatures (HT-ESEM). Actually, the use of HT-ESEM allowed the in situ observation of the samples during long term heat treatments up to 1400 °C under various atmospheres (Podor *et al.*, 2012). In a first step, lanthanide and actinide oxide grains with spherical shapes were synthesized by means of soft chemistry (Nkou Bouala *et al.*, 2014), in order to investigate chemical systems with shapes close to those used in models. Then the HT-ESEM was used to investigate the first stage of sintering on three different systems (a single grain and two and three grains in contact) in the temperature range of 900–1200 °C:

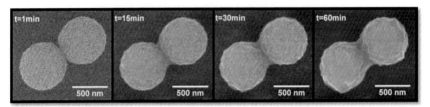

**Figure 24** In situ HT-ESEM micrographs series of two CeO$_2$ nanospheres showing the evolution of neck size and dihedral angles between the grains at T = 1100 °C.

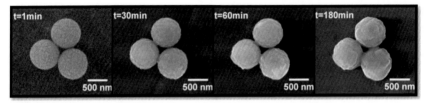

**Figure 25** In situ HT-ESEM micrographs series of three CeO$_2$ nanospheres showing the evolution of neck size, dihedral angles and porosity between the grains at T = 1100 °C.

- Monitoring of a single grain led to the evolution of the number of crystallites included in the sphere. From the micrographs series, the time necessary to reach a spherical single crystal was determined, as well as the activation energy necessary for the growth of crystallites.
- The observation of the morphological modifications of two- and three-grain arrangements led to assess the evolution of several parameters of interest, such as neck size, dihedral angles between the spheres, and distance between the grain centers. From the micrographs series, it was possible to identify experimentally for the first time the mechanisms of neck growth between the grains (Figures 24–25).

The use of HT-ESEM observations appears to be of great interest for the study of sintering phenomena. The exploitation of micrograph series allows for determining original and fundamental experimental data and characteristics of the processes occurring during the initial stage of sintering.

## REFERENCES

Bernache-Assolant, D. (1993). Chimie-physique du frittage. *Hermes Eds*, 348.

Nkou Bouala, G.I., Clavier, N., Podor, R., Cambedouzou, J., Mesbah, A., Brau, H.P., et al. (2014). Preparation and characterisation of uranium oxides with spherical shape and hierarchical structure. *CrystEngComm, 16*, 6944–6954.

Podor, R., Clavier, N., Ravaux, J., Claparède, L., Dacheux, N., & Bernache-Assollant, D. (2012). Dynamic aspects of cerium dioxide sintering: HT-ESEM study of grain growth and pore elimination. *Journal of the European Ceramic Society, 32,* 353–362.

Qin, J., Yang, R., Liu, G., Li, M., & Shi, Y. (2010). Grain growth and microstructural evolution of yttrium aluminium garnet nanocrystallites during calcination process. *Materials Research Bulletin, 45,* 1426–1432.

Slamovich, E.B., & Lange, F.F. (1990). Densification behavior of single-crystal and polycrystalline spherical particles of zirconia. *Journal of the American Ceramic Society, 73*(11), 3369–3375.

Wakai, F. (2006). Modeling and simulation of elementary process in ideal sintering, *Journal of the American Ceramic Society, 89*(5), 1471–1484.

# In Situ Transmission Electron Microscopy of High-Temperature Phase Transitions in Ge-Sb-Te Alloys

**Katja Berlin\*, Achim Trampert**
Paul-Drude-Institut für Festkörperelektronik, Hausvogteiplatz 5-7, 10117 Berlin, Germany
\*Corresponding author: e-mail address: berlin@pdi-berlin.de

## INTRODUCTION

The direct observation of atomic processes during structural phase transitions or grain boundary motion is of fundamental importance for understanding the properties of phase change materials. In situ transmission electron microscopy (TEM) enables excellent spatial and analytical resolution allowing the acquisition of data about the atomic structure in combination with heating experiments. The system under study is the ternary Ge-Sb-Te alloy, which is characterized by an amorphous state at room temperature and phase transitions to cubic (about 150–165 °C) (Seo et al., 2000; Friedrich et al., 2000) and hexagonal (about 340–360 °C) (Friedrich et al., 2000; Kooi et al., 2004) crystal structures at higher temperatures.

## EXPERIMENTAL DETAILS

The Ge-Sb-Te (GST) thin films were deposited on a cleaned Si-(111) substrate in an ultra-high vacuum chamber by using effusion cells for the three elements. One sample was prepared at room temperature and subsequently heated to form a polycrystalline film, the other was epitaxially grown on Si (Rodenbach et al., 2012). Cross-sectional TEM samples were prepared using the standard procedure of mechanical grinding, dimpling and ion beam milling. In situ TEM is performed with a JEOL 3010 operated at 200 kV using a JEOL double-tilt specimen heating holder and controller unit. The temperatures stated here referred to the temperature given by the controller unit. The temporal resolution of the recorded videos are 0.04 s.

## IN SITU TEM IMAGING
### 1 Grain Boundary Dynamics and Phase Transition

A thin polycrystalline $Ge_1Sb_2Te_4$ film (composition confirmed by energy electron loss spectroscopy) is annealed to 400 °C for in situ observation of the motion of grain boundaries. Dark-field imaging conditions are used for achieving an accurate determination of the grain boundary position. Figure 26 shows a sequence of dark-field snapshots taken from a video, where the bright grain is characterized by a preferential orientation to the substrate: hexagonal basal planes are parallel to Si-(111) planes. During the grain boundary motion, the preferential grain grows as measured by the distance $d_0$ between the actual boundary position and an obstacle at the interface versus time. The corresponding graph in Figure 27 reveals a steplike character instead of a continuously increasing progression. This behavior could be caused by the cubic-to-hexagonal phase transition accompanying the grain boundary motion with the formation of the complex layered structure (Kooi *et al.*, 2002) of hexagonal $Ge_1Sb_2Te_4$.

## 2 Step-Flow Motion

Another important aspect of phase transitions addresses the crystallization dynamics and the evolution of the growth front morphology. An epitaxially aligned cubic $Ge_2Sb_2Te_5$ film (Rodenbach *et al.*, 2012) is annealed at about 400 °C. At the interface region, unexpected amorphization and recrystallization is observed under $e^-$-beam irradiation. The crystallization proceeds along the highly stepped surface similar to the step-flow growth mode during epitaxy on vicinal surfaces (Figure 27). However, within our temporal resolution, the multiple steps grow both block by block and in a two-layer

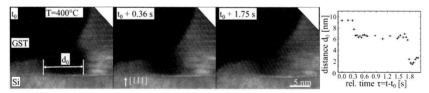

**Figure 26** Dark-field TEM images taken from a video sequence showing the recrystallization of GST thin film at 400 °C: The grain with the preferred orientation of basal planes parallel to Si (111) planes grows laterally. The graph shows the result of the growth analysis: a stepwise and fast growth process.

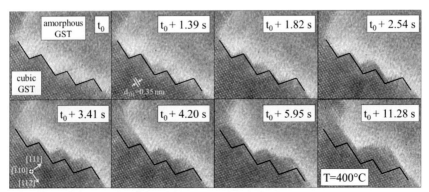

**Figure 27** HRTEM image series taken at 400 °C showing the crystallization of cubic GST from an amorphous reservoir. The initially growth front from the first image (at t0) is repeated as a black line in the following images (t>0) for reference.

growth, rather than by simple atom migration. This behavior could again be caused by the complex structure. In order to get the right composition, the material needs at least two layers to replicate the surrounding matrix.

## REFERENCES

Friedrich, I., Weidenhof, V., Njoroge, W., Franz, P., & Wuttig, M. (2000). Structural transformations of Ge2Sb2Te5 films studied by electrical resistance measurements. *Journal of Applied Physics, 87,* 4130–4134.

Kooi, B.J., Groot, W.M.G., & De Hosson, J.T.M. (2004). In situ transmission electron microscopy study of the crystallization of Ge2Sb2Te5. *Journal of Applied Physics, 95,* 924–932.

Rodenbach, P., Calarco, R., Perumal, K., Katmis, F., Hanke, M., Proessdorf, A., et al. (2012). Epitaxial phase-change materials. *Physica Status Solidi RRL, 6,* 415–417.

Kooi, B.J., & De Hosson, J.T.M. (2002). Electron diffraction and high-resolution transmission electron microscopy of the high temperature crystal structures of Ge_xSb_2Te_3+x (x = 1,2,3) phase change material. *Journal of Applied Physics, 92,* 3584–3590.

Seo, H., Jeong, T. H, Park, J.W., Yeon, C., Kim, S.J., & Kim S.Y. (2000). Investigation of crystallization behavior of sputter-deposited nitrogen-doped amorphous Ge2Sb2Te5 thin films. *Japanese Journal of Applied Physics, 39,* 745–751.

# In Situ Tem of Catalytic Nanoparticles

# Correlative Microscopy for In Situ Characterization of Catalyst Nanoparticles Under Reactive Environments

**Renu Sharma\***

Center for Nanoscale Science and Technology, National Institute of Standards and Technology, Gaithersburg, MD 20899–6203

\*Corresponding author: e-mail address: renu.sharma@nist.gov

In recent years, the environmental transmission scanning electron microscope (ESTEM) has been successfully employed to elucidate the structural and chemical changes occurring in the catalyst nanoparticles under reactive environments. While atomic–resolution images and the combination of high spatial and energy resolution is ideally suited to distinguish between active and inactive catalyst particles and identify active surfaces for gas adsorption, unambiguous data can only be obtained from the area under observation. This lack of global information available from TEM measurements is generally compensated for by using other, ensemble measurement techniques such as X-ray or neutron diffraction, X-ray photoelectron spectroscopy, infrared spectroscopy, and Raman spectroscopy. However, it is almost impossible to create identical experimental conditions in two separate instruments to make measurements that can be directly compared. Moreover, ambiguities in ESTEM studies may arise from the unknown effects of the incident electron beam and uncertainty of the sample temperature. We have designed and built a unique platform that allows us to concurrently measure atomic-scale and micro-scale changes occurring in samples subjected to identical reactive environmental conditions by incorporating a Raman spectrometer on the ESTEM. We have used this correlative microscopy platform (1) to measure the temperature from a 60-$\mu m^2$ area using Raman shifts, (2) to investigate light-matter interactions, and (3) as a heating source for concurrent optical and electron spectroscopy, such as cathodoluminescence, electron energy loss spectroscopy (EELS), and Raman. Details of the design, function, and capabilities will be illustrated with results obtained from in situ combinatorial measurements.

# Applications of Environmental TEM for Catalysis Research

Jakob B. Wagner*, Christian D. Damsgaard, Filippo Cavalca**,
Elisabetta M. Fiordaliso, Diego Gardini, Thomas W. Hansen
Center for Electron Nanoscopy, Technical University of Denmark, Fysikvej Building 307, DK-2800 Kgs. Lyngby, Denmark
*Corresponding author: e-mail address: jakob.wagner@cen.dtu.dk

In the quest for a better understanding of the dynamics of heterogeneous catalysts under working conditions, in situ characterization plays a key role. Catalyst efficiency and sustainability are strongly linked to the dynamic morphology and atomic arrangements of the active entities of these complex materials. Here, we provide examples of how environmental transmission electron microscopy (ETEM) can be applied in catalysis research.

ETEM is a unique tool combining imaging and spectroscopy capabilities with a spatial resolution approaching 1 Å at elevated temperatures and controlled gaseous environments. As with any other characterization technique, ETEM has drawbacks. Limitations in sample geometry and field of view call for use of complementary in situ characterization techniques. Such techniques could be in situ X-ray based techniques such as X-ray diffractometer (XRD), X-ray absorption spectroscopy (XAS), and extended X-ray absorption fine structure (EXAFS). The combination of techniques will paint a more complete picture of the sample under a reactive environment.

In order to take full advantage of the complementary in situ techniques, transfer under reaction conditions is essential. Here, we introduce the in situ transfer concept by use of a dedicated TEM transfer holder capable of enclosing the sample in a gaseous environment at temperatures up to approx. 900 °C. The holder is compatible with other in situ technique set-ups (Damsgaard et al., 2014).

Figure 28 shows the concept of the transfer holder. In this experiment, cuprous oxide has been initially characterized in the ETEM at room temperature in a vacuum, then transferred to an in situ XRD setup, reduced in hydrogen, and finally transferred back to the microscope in $10^5$ Pa of $H_2$ at 220 °C.

In general, the catalytic behavior of nanoparticle systems can be greatly influenced by the synthesis procedure used. The different stages of the

---

**Present address: Haldor Topsøe A/S, 2800 Kgs. Lyngby, Denmark

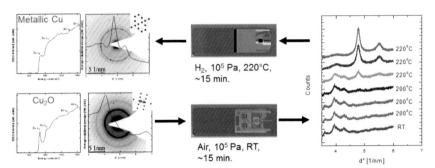

**Figure 28** In situ transfer of copper-containing nanoparticles between ETEM and in situ XRD setup. The cuprous oxide particles are reduced in the XRD and transferred under an elevated temperature in hydrogen to the ETEM, keeping the copper metallic.

formation process, such as drying, calcination, and reduction, can all be optimized by tweaking parameters such as temperature, time of treatment, and gaseous environment. Following the process of identified volume subsets of the catalyst material by means of electron beam–based characterization at the different stages of synthesis gives insight in the formation mechanisms of individual nanoparticles, from precursor to active catalyst.

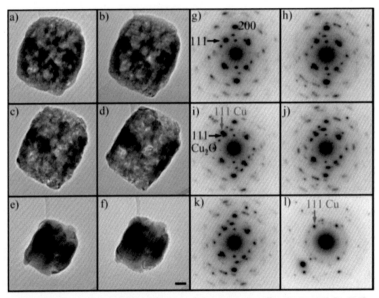

**Figure 29** Dynamic study of degradation of $Cu_2O$ to metallic Cu in 500 Pa $H_2O$ under irradiation of visible light in situ in the ETEM. (a)–(f) TEM imaging. Scale bar 50 nm. (g)–(l) Corresponding selected area diffraction patterns. These images and diffraction patterns are acquired in the absence of water and light illumination. Each frame separated by 15 min. *Figure partly adapted from Cavalca et al. (2013).*

Converting solar energy into chemical bonds, and thereby increasing the storage capability of light energy harvesting, require the use of photo-catalysts. In order to characterize and study such catalysts under relevant conditions, we have developed a TEM holder capable of exposing a sample to visible light inside the microscope. In this way, photo induced reactions and phenomena such as $Cu_2O$ degradation (Figure 29) are studied in situ under different gaseous environments (Cavalca et al., 2013).

## REFERENCES

Cavalca, F., Laursen, A.B., Wagner, J.B., Damsgaard, C.D., Chorkendorff, I., & Hansen, T.W. (2013). Light-induced reduction of cuprous oxide in an environmental transmission electron microscope. *ChemCatChem*, *5*, 2667–2672.

Damsgaard, C.D., Zandbergen, H., Hansen, T.W., Chorkendorff, I., & Wagner, J.B. (2014). Controlled environment specimen transfer. *Microscopy and Microanalysis*, *20*, 1038–1045.

# Selected Posters

# Gold Nanoparticle Movement in Liquid Investigated by Scanning Transmission Electron Microscopy

**Marina Pfaff[a],\*, Niels de Jonge[a,b,c]**

[a]INM-Leibniz Institute for New Materials, Campus D2 2, 66123 Saarbrücken, Germany
[b]Department of Physics, University of Saarland, Campus A5 1, 66123 Saarbrücken, Germany
[c]Department of Molecular Physiology and Biophysics, Vanderbilt University School of Medicine, 2215 Garland Ave, Nashville, TN 37232-0615
\*Corresponding author: e-mail address: marina.pfaff@inm-gmbh.de

We have used two experimental approaches to study the motion of gold nanoparticles in liquid: scanning transmission electron microscopy in liquid (liquid STEM), and environmental scanning electron microscopy (ESEM) equipped with a wet-STEM detector (de Jonge & Ross, 2011; Peckys & de Jonge, 2014). In liquid STEM, a microfluidic chamber assembled from two Si microchips with electron-transparent SiN windows is used to enclose the nanoparticle solution, and separate it from the vacuum in the microscope chamber (Figure 30A). The images were recorded at 200 keV in STEM mode using the annular-dark-field (ADF) detector providing Z-contrast. With this method, the particles show a high contrast compared to the

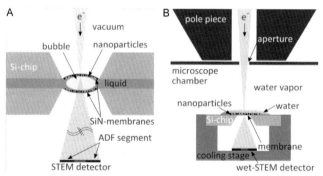

**Figure 30** Principles of liquid STEM and ESEM STEM imaging of nanoparticles in liquid. In both modes, an image is obtained by scanning a focused electron beam over the sample and detecting the transmitted electrons for each pixel. (a) In liquid STEM, the nanoparticle solution is enclosed between two SiN windows. Depending on the imaging conditions, bubbles can be formed under electron irradiation. (b) In ESEM STEM, the nanoparticles are imaged in a thin liquid layer on a SiN membrane. To maintain a stable liquid layer, the pressure in the specimen chamber and the temperature of the cooling stage are adjusted.

background signal originating from the liquid layer. A spatial resolution of 1.5 nm was achieved for gold nanoparticles (Ring & de Jonge, 2012). In comparison to liquid STEM, there is no need of a microfluidic chamber in ESEM STEM (Figure 31B). The nanoparticles were imaged in a droplet of water on an SiN window that was cooled to 3 °C. To maintain a stable liquid layer thickness, the water vapor pressure in the microscope chamber is adjusted to 750 Pa. To image the particles, we used 30 keV, resulting in a decreased but still nanoscale resolution compared to STEM at 200 keV.

For the liquid STEM experiments, 30-nm-diameter gold nanoparticles were investigated in a glycerol solution with 20% water. This solution has a higher viscosity compared to pure water and therefore slows the nanoparticle movement. Nanoparticles stabilized by a coating of positively charged thiolated chitosan (TCHIT) were used. The electron irradiation induced the detachment of the particles from the SiN membrane, and we observed a subsequent movement and agglomeration of the particles (Figure 31). A bubble was formed during imaging by the interaction of the liquid with the electron beam. As a result, the liquid layer thickness in the bubble region was decreased, and we observed that the velocity of the nanoparticles depended on the thickness of the liquid layer. The thinner the liquid layer, the slower the particles move. The calculated theoretical displacement due to Brownian motion was more than two orders of magnitude higher than our experimental data. In the ESEM STEM experiments, we imaged 30-nm, citrate-stabilized nanoparticles in pure water. Two different regimes of velocities were observed, very fast ones and much slower ones. Interestingly, even the velocities of the fastest nanoparticles were much smaller than expected on the basis of Brownian motion.

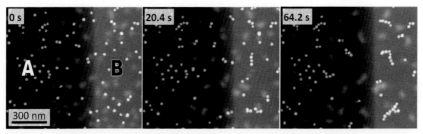

**Figure 31** ADF liquid STEM time series of Au-nanoparticles in a glycerol/water solution. A bubble was formed due to electron irradiation (region A), reducing the liquid layer thickness. The particles in the continous liquid layer (region B) moved and agglomerated faster than in the bubble area. (E = 200 keV, $I_{beam}$ = 180 pA, pixel time: 1 $\mu s$).

Our observations point toward the presence of interactions in addition to Brownian motion slowing down the motion of nanoparticles in liquid. Possible explanations could involve van der Waals forces or electrostatic interactions between the gold nanoparticles and the SiN window or the surface of the droplet (Ring & de Jonge, 2012). The reduced velocities compared to Brownian motion are beneficial for electron microscopical investigations of particles in liquid since nanoscale resolution is not much degraded by the movement of the nanoparticles within the image acquisition.

## REFERENCES

de Jonge, N., & Ross, F.M. (2011). Electron microscopy of specimens in liquid. *Nature Nanotechnology, 6,* 695–704.

Peckys, D.B., & de Jonge, N. (2014). Liquid scanning transmission electron microscopy: Imaging protein complexes in their native environment in whole eukaryotic cells. *Microscopy and Microanalysis, 20,* 346–365.

Ring, E.A., & de Jonge, N. (2012). Video-frequency scanning transmission electron microscopy of moving gold nanoparticles in liquid. *Micron, 43,* 1078–1084.

# Correlating Scattering and Imaging Techniques: In Situ Characterization of Gold Nanoparticles Using Conventional TEM

**Dimitri Vanhecke\*, Benjamin Michen, Sandor Balog, Christoph Geers, Carola Endes, Barbara Rothen-Rutishauser, Alke Petri-Fink**
BioNanomaterials Group, Adolphe Merkle Institute, University of Fribourg, Switzerland
\*Corresponding author: e-mail address: dimitri.vanhecke@unifr.ch

Nanomaterials have been promised a great future and are already being used in various applications. The recently updated definition of the term *nanomaterial*, to be used in all European Union (EU) legislation, is "A nanomaterial means a natural, incidental, or manufactured material containing particles, in an unbound state or as an aggregate or as an agglomerate and where, for 50% or more of the particles in the number size distribution, one or more external dimensions is in the size range 1 nm–100 nm." In this definition, a nanoparticle has all three dimensions in the nanoscale. Because their physical and chemical properties depend very much on their size and shape, their characterization is of utmost importance. An in situ observation technique such as dynamic light scattering (DLS) is commonly used to probe the dispersion state of colloids dispersed in a fluid. This technique is very precise in determining mean particle size and its distribution for monomodal particle samples. But when multimodal distributions evolve (for example, induced by aggregation), the quantification becomes challenging due to the assumptions made in the applied models. Unfortunately, the median of the number-weighted size distribution, which has become an instrumental parameter in nanomaterial's legislation, cannot be retrieved by scattering methods. Transmission electron microscopic (TEM) analysis can provide model-free data and the number size distribution. However, in order to obtain TEM samples of nanoparticles in a fluid, the solvent (mostly water) must be removed. Drying effects will take place during sample preparation, which leads to the accumulation and aggregation and hence introduces bias or artifacts in the analysis (Michen *et al.*, 2014). It is impossible to differentiate nanoparticle aggregates caused by drying effects from aggregates that were formed prior to drying (e.g., due to colloidal instability of the system). Thus, TEM has long been thought to be unsuitable for quantitative investigations of nanoparticles and colloidal aggregates of nanoparticles in liquids. Here, we present a straightforward, cost-effective protocol that circumvents

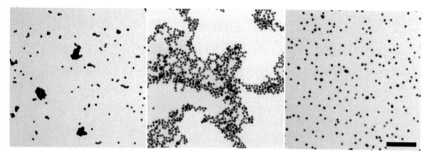

**Figure 32** The effect of aggregation induced during sample preparation (drying) can be overcome by colloidal stabilization by macromolecular agents. From left to right: increasing BSA concentration on the same stock solution of 15-nm Au particles. Without colloidal stabilization, strong aggregation form (left). Suboptimal colloidal stabilization drives particles together in loose aggregations by a phenomenon known as polymer bridging (middle). Sufficient stabilization prevents aggregation and yields micrographs of separate, single particles; an ideal target for automated size estimation (left). Bar = 200 nm.

the drying issue (Figure 32) by stabilizing the colloidal system using a macromolecular agent bovine serum albumin (BSA). TEM samples with various ratios of proteins to nanoparticles have been prepared and resulted in different spatial distributions of nanoparticles on the TEM grid. When a sufficient high-protein concentration was used, the resulting TEM images allow for automated data collection on a large number of nanoparticles which is in very good agreement with data obtained from dynamic light scattering and thus the actual dispersion state.

## REFERENCE

Michen, B., Balog, S., Rothen-Rutishauser, B., Petri-Fink, A., & Vanhecke, D. (2014). TEM sample preparation of nanoparticles in suspensions: understanding the formation of drying artefacts. *Imaging & Microscopy, 6*(3), 39–41.

# The Effects of Salt Concentrations and pH on the Stability of Gold Nanoparticles in Liquid Cell STEM Experiments

**Andreas Verch[a],\*, Justus Hermannsdörfer[a], Marina Pfaff[a], Niels de Jonge[a,b,c]**

[a]INM-Leibniz Institute for New Materials, Campus D2 2, 66123 Saarbrücken, Germany
[b]Department of Molecular Physiology and Biophysics, Vanderbilt University School of Medicine, 2215 Garland Ave, Nashville, TN 37232-0615
[c]Department of Physics, University of Saarland, Campus A5 1, 66123 Saarbrücken, Germany
\*Corresponding author: e-mail address: andreas.verch@inm-gmbh.de

With the introduction of commercially available transmission electron microscopy (TEM) holders for liquid specimens in recent years, the interest in liquid-cell TEM has greatly risen. Nowadays, this method is applied in scientific fields as different as battery research, cell biology, and materials science in order to gain information about structures and processes on the nanoscale in a liquid or the natural environment (de Jonge & Ross, 2011).

However, the presence of liquids and, in most cases, also solutes dramatically increases the number of chemical reactions conceivable during an experiment in the electron microscope compared to TEM in a vacuum. In addition to interactions with solid materials, as known from conventional electron microscopy experiments, the electron beam influences the chemistry of the liquid. Electrons penetrating through the liquid excite electronic states in molecules or atoms or remove electrons from the electron shell, resulting in the generation of highly reactive, transient species (Henglein et al., 1969). Some of these compounds are strong-reducing (e.g., solvated electrons) others are powerful oxidizing agents (e.g., hydroxyl radicals), that are capable of dissolving metals as noble as gold. These by-products are often not desired, as they may have severe consequences on the outcome of liquid cell transmission electron microscopy experiments, contingently leading to entirely unexpected results. Hence, a better understanding of these electron beam–induced processes, and avoiding them is needed in order to design experiments comparable to those without electron beam exposure.

In this study, we focused on the impact of the electron beam on gold nanoparticles in an aqueous environment. We utilized TEM experiments in scanning mode (e.g., STEM) in order to obtain a good handle on the dose introduced into the observed sample area. By varying the properties of the solution (e.g., the salt concentration or pH) and the imaging conditions, we

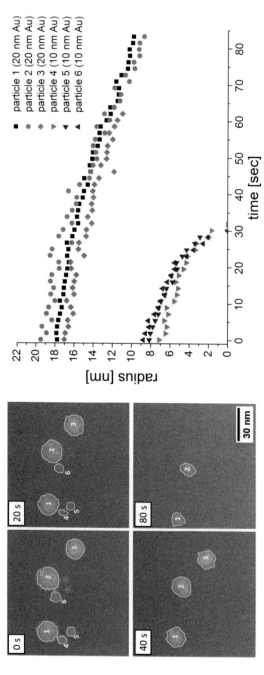

**Figure 33** STEM experiments of gold nanoparticles in liquid acquired at a 200-keV beam energy. (A) Time series showing the dissolution of gold nanoparticles at a magnification of 1,000,000 x. Yellow borders indicate the areas used for the analysis of dissolution kinetics presented in (B). (B) Evolution of the particle radii.

altered the composition of the transient species generated by the electron beam (Schneider *et al.*, 2014). It became apparent that the addition of high concentrations of chloride ions, destabilized gold nanoparticles. Moreover, the size of the nanoparticles influenced the susceptibility toward dissolution, with smaller particles dissolving faster than bigger ones (Figure 33).

## ACKNOWLEDGMENTS

We thank D.B. Peckys for help with the experiments, and Protochips Inc, NC for providing the microchips and liquid specimen holder. We thank E. Arzt for his support through INM. This research was in part supported by the Leibniz Competition 2014.

## REFERENCES

de Jonge, N., & Ross, F.M. (2011). Electron microscopy of specimens in liquid. *Nature Nanotechnology*, *6*, 695–704.

Henglein, A., Schnabel, W., & Wendenburg, J. (1969). *Einführung in die Strahlenchemie mit praktischen Anleitungen*. Weinheim, Germany, Verlag Chemie GmbH.

Schneider, N.M., Norton, M.M., Mendel, B.J., Grogan, J.M., Ross, F.M., & Bau, H.H. (2014). Electron-water interactions and implications for liquid cell electron microscopy. *Journal of Physical Chemistry C*, *118*, 22373–22382.

# Bridging the Gap Between Electrochemistry and Microscopy: Electrochemical IL-TEM and In Situ Electrochemical TEM Study

Nejc Hodnik*, Claudio Baldizzone, Jeyabharathi Chinnaya, Gerhard Dehm, Karl Mayrhofer

Max-Planck-Institut für Eisenforschung, Max-Planck-Str. 1 40237 Düsseldorf, Germany
*Corresponding author: e-mail address: n.hodnik@mpie.de

Generally, in order to make the water and carbon energy cycles economically competitive in terms of today's fossil and nuclear energy, energy conversion materials have to be further optimized and therefore studied in more detail. More specifically, the fundamental details of platinum–based proton exchange membrane fuel cell (PEM–FC) catalyst degradation are still not clear (Mayrhofer et al., 2008; Meier et al., 2012). Missing insights can only be gained with an introduction of new advance characterization techniques like identical location transmission electron microscopy (IL–TEM) and in situ electrochemical TEM. A generally accepted technique, besides electrochemistry, used to study the structure of nanoparticles is a vacuum TEM equipped with numerous detectors enabling many characterization methods, such as electron diffraction, high-resolution TEM (HR–TEM), scanning TEM (STEM), and energy dispersive spectroscopy (EDS), which provide a complete description of studied nanomaterial characteristics. This information is commonly used to interpret nanoparticle activity and stability. However, the concerns still remain whether the surface structure and morphology change of random spots is representative for the whole sample and if the structure and morphology integrity get disrupted when nanoparticles come in contact with the atmosphere or get immersed in the liquid electrolyte. It can reconstruct or deconstruct the metal surface depending on the interaction energy of thermodynamic driving forces. It can also change unevenly across the sample. In order to access the information of real surface structure at relevant conditions, identical locations and, more important, in situ observations are mandatory. Before IL–TEM was the only technique showing degradation on the same nanoparticle (Hodnik et al., 2014; Mayrhofer et al., 2008; Meier et al., 2012). Combination of this approach with conventional electrochemistry provides a complete description of the catalytic activity and stability in relation to the surface structural characteristics.

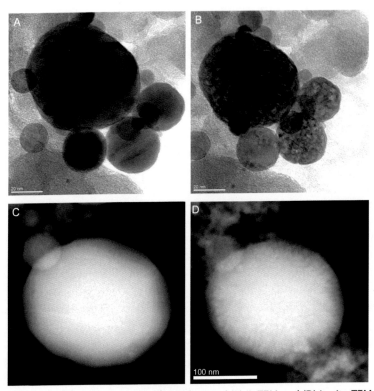

**Figure 34** (A) IL-TEM and (C) in situ TEM before and (B) IL-TEM and (D) in situ TEM after dealloying treatment of PtCu$_3$ nanparticle in 0.1 M HClO$_4$. IL-TEM images were already published in Hodnik et al. (2014). *Reproduced by permission of the PCCP Owner Societies.*

In this study, we present dealloying of polydisperse PtCu$_3$ nanoparticles. IL-TEM images (Figures 34A and B) show that particles larger than 20 nm form a porous structure upon a potential hold experiment (1.2 V vs RHE for 2 h in 0.1 M HClO$_4$) (Hodnik *et al.*, 2014). We provide the first preliminary results done with electrochemical in situ TEM (Poseidon 500, Protochips) confirming the pore formation upon electrochemical biasing (Figure 34C and D).

## ACKNOWLEDGMENTS

Nejc Hodnik would like to acknowledge the FP7-PEOPLE-2013-IEF Marie Curie Intra-European Fellowship (project 625462–ElWBinsTEM).

# REFERENCES

Hodnik, N., Jeyabharathi, C., Meier, J.C., Kostka, A., Phani, K.L., Recnik, A., et al. (2014). Effect of ordering of PtCu3 nanoparticle structure on the activity and stability for the oxygen reduction reaction. *Physical Chemistry Chemical Physics, 16,* 13610–13615.

Mayrhofer, K.J.J., Meier, J.C., Ashton, S.J., Wiberg, G.K.H., Kraus, F., Hanzlik, M., & Arenz, M. (2008). Fuel cell catalyst degradation on the nanoscale. *Electrochemistry Communications, 10,* 1144–1147.

Meier, J.C., Galeano, C., Katsounaros, I., Topalov, A.A., Kostka, A., Schüth, F., & Mayrhofer, K.J.J. (2012). Degradation mechanisms of Pt/C fuel cell catalysts under simulated start-stop conditions. *ACS Catalysis, 2,* 832–843.

# Using a Combined TEM/Fluorescence Microscope to Investigate Electron Beam–Induced Effects on Fluorescent Dyes Mixed into an Ionic Liquid

**E. Jensen[a,*], S. Canepa[a], R. Liebrechts[b], K. Mølhave[a]**
[a]DTU Nanotech, Ørsteds Plads, Building 345E, DK-2800 Kongens Lyngby
[b]Department of Biomedical Sciences, Endocrinology Research Section, Blegdamsvej 3, DK-2200 Copenhagen
*Corresponding author: e-mail address: eric.jensen@nanotech.dtu.dk

The implementation of in situ liquid-phase electron microscopy has increased significantly in the last decade (Ross and de Jonge, 2011). Several chemical reactions, with some induced by the electron beam, have been observed in situ (Radisic et al., 2006; Williamson et al., 2003; Zheng et al., 2009), but the effects of the electron beam on these experiments is still not fully understood (Abellan et al., 2014; Evans et al., 2011; Grogan et al., 2014; Schneider et al., 2014).

The electron beam will generate redox active species, as well as local changes in pH, and in turn affect the local chemical environment being imaged. In this study, we tested the influence of irradiation on the fluorescence signal from a liquid sample in the TEM (Figure 35). The combined TEM/fluorescence microscope imaging modes offered by the FEI Technai iCorr was used to image a vacuum compatible ionic liquid (1-butyl-3-methylimidazolium tetrafluoroborate) mixed with two fluorescent dyes: fluorescein and Nile blue. The two dyes were mixed into the ionic liquid such that the fluorescent signal could be imaged with the fluorescence microscope.

The investigation showed varied results. In the fluorescein sample (Figure 35A), the fluorescent signal was increased after electron exposure. The ionic liquid became polymerized, as evidenced by the irradiated areas not gradually becoming diffuse after electron exposure. For the Nile blue sample, no such evidence was found. Rather, the fluorescent signal deteriorated over time, which indicates that the reaction stabilized or that the fluorescent dye was replenished by diffusion.

When performing experiments at different dose rates, a nearly linear correlation between intensity loss and dose rate was observed for the fluorescein sample, but the Nile blue showed no correlation.

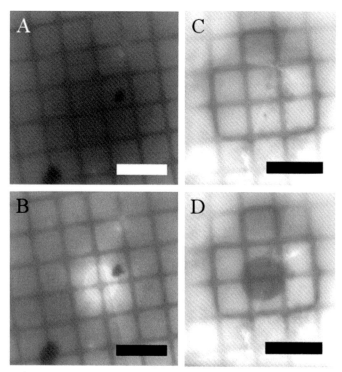

**Figure 35** Fluorescense-microscopy images of TEM grids with ionic liquid and fluorescent dye. (A) and (C) are before exposure to the electron beam and (B) and (D) are afterward. The TEM grids in (A) and (B) are covered with ionic liquid mixed with 550 $\mu$M fluorescein, and the ones in (C) and (D) are covered with ionic liquid mixed with 93-mM of Nile blue. There is a clear fluorescence change from the circular central region's exposure to a dose rate of 12 e·Å$^{-2}$·s$^{-1}$. The scale bar in all images is 15 $\mu$m.

These initial experiments (Figure 36) clearly showed a strong change in fluorescence after exposure to low doses of the electron beam. In order to fully quantify these results, it will be necessary to further control the thickness of the ionic liquid.

## REFERENCES

Abellan, P., Mehdi, B.L., Parent, L.R., Gu, M., Park, C., Xu, W., et al. (2014). Probing the degradation mechanisms in electrolyte solutions for L-ion batteries by in situ transmission electron microscopy. *Nano Letters, 14*(3), 1293–1299.

Evans, J.E., Jungjohann, K.L., Browning, N.D., & Arslan, I. (2011). Controlled growth of nanoparticles from solution with in situ liquid transmission electron microscopy. *Nano Letters, 11*(7), 2809–2813.

Grogan, J.M., Schneider, N.M., Ross, F.M., & Bau, H.H. (2014). Bubble and pattern formation in liquid induced by an electron beam. *Nano Letters, 14*(1), 359–364.

Radisic, A., Vereecken, P.M., Hannon, J.B., Searson, P.C., & Ross, F.M. (2006). Quantifying electrochemical nucleation and growth of nanoscale clusters using real-time kinetic
data. *Nano Letters*, *6*(2), 238–242.

Ross, F.M., & de Jonge, N. (2011). Electron microscopy of specimens in liquid. *Nature Nanotechnology*, *6*(11), 695–704.

Schneider, N.M., Norton, M.M., Mendel, B.J., Grogan, J.M., Ross, F.M., & Bau, H.H.
(2014). Electron-water interactions and implications for liquid cell electron microscopy.
*Journal of Physical Chemistry*, *118*(38), 22373–22382.

Williamson, M.J., Tromp, R.M., Vereecken, P.M., Hull, R., & Ross, F.M. (2003). Dynamic
micoscopy of nanoscale cluster growth at the solid-liquid interface. *Nature Materials*, *2*(8),
532–536.

Zheng, H., Claridge, S., Minor, A.M., Alivisatos, P., & Dahmen, U. (2009). Nanocrystal
diffusion in a liquid thin film observed by in situ transmission electron microscopy. *Nano
Letters*, *9*(6), 2460–2465.

# Microfabricated Low-Thermal Mass Chips System for Ultra-Fast Temperature Recording During Plunge freezing for Cryofixation

S. Laganá*, T. Kjøller Nellemann, E. Jensen, K. Mølhave

DTU Nanotech, Ørsteds Plads B345Ø, DK-2800 Kgs. Lyngby

*Corresponding author: e-mail address: simla@nanotech.dtu.dk

Plunge freezing is commonly used as a cryofixation technique for the stabilization of biological materials prior to electron microscopy (EM) (Dobro et al., 2010; Milne et al., 2013). However, even if plunge freezing is the easiest way to retain the original structure of the sample, it is very difficult to achieve the high cooling rates (Ryan et al., 1992) necessary for obtaining amorphous ice and thus avoiding freezing artefacts such as ice crystal distortions. We present the development of a low-mass, microchip–based system for the measurement of ultra-fast freezing rates during cryo–immobilization of small hydrated volumes such as living cells and thin liquid films. Our system enables the change in temperature to be tracked accurately over the short timescales involved, by measuring the resistance of metallic thin–film resistors lithographically defined on the membrane. The chip system was used to determine cooling rates in slush nitrogen, propane and a mixture of propane and ethane with and without liquid samples. We measured cooling rates of over 100.000 K/s (see Figure 36) and heat transfer coefficients over 10.000 W/m$^2$K on the dry chip, with lower values when liquid layers were present, and occasionally even higher values in slush nitrogen.

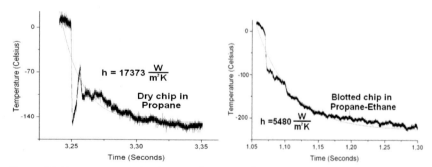

**Figure 36** Temperature reduction as a function of time on the membrane (200 nm of silicon-rich nitride) of (A) a dry chip with a gold wire contact on top dropped in propane and (B) a wet-blotted chip in propane-ethane.

The microfabricated systems point to ways of making recordings of individual sample freezing with sub-millisecond time-resolution processes in liquids with subsequent electron microscopic imaging of the sample on the thin membranes.

## REFERENCES

Dobro, M.J., Melanson, L., Jensen, G.J., & McDowall, A.W. (2010). Plunge freezing for electron cryomicroscopy. In *Methods in Enzymology*. Elsevier Inc. doi:10.1016/ S0076-6879(10)81003-1.
Milne, J.L.S., et al. (2013). Cryo-electron microscopy—A primer for the nonmicroscopist. *FEBS Journal, 280,* 28–45. doi:10.1111/febs.12078.
Ryan, K.P. (1992). Cryofixation of Tissues for Electron-Microscopy *Scanning Microscopy*, 6, 715–743.

# In Situ SEM Cell for Analysis of Electroplating and Dissolution of Cu

**R. Møller-Nilsen[a,*], S. Colding-Jørgensen[a], E. Jensen[a], M. Arenz[b], K. Mølhave[a]**
[a]DTU Nanotech, Ørsteds Plads B345Ø, DK-2800 Kgs. Lyngby
[b]Copenhagen University, Department of Chemistry, Universitetsparken 5, DK-2100 CPH Ø
*Corresponding author: e-mail address: rerom@nanotech.dtu.dk

Electroplating of copper is a widely used technique for aesthetic coatings, creating electrodes, and precoating prior to additional electroplating. Recently it has become possible to directly see the growth on a nanometer scale. Theory on the nucleation density laid down by Hyde and Compton (2003) was augmented by Radisic et al. (2006a, 2006b) after they found by means of in situ TEM that the nucleation density was three orders of magnitude more than predicted. More recently, Schneider et al. (2014) studied the effect of depletion on growth morphology.

While Radisic et al. (2006a, b). used a TEM chip with a very limited volume and looking through 1–2 $\mu$m of electrolyte, we are using an in situ SEM setup that allows us to have an electrolyte volume on the order of 1 mL (Jensen et al., 2013). A larger total volume and more open liquid cell geometry ensures a better diffusive supply from the bulk, which means that the electrolyte close to the electrode remains more representative of the bulk for a longer duration of electron radiation and electrochemical depletion.

In addition to studying the morphology in different deposition current regimes, we investigated how the electron beam affects the experiment by blanking the beam during deposition and dissolution and relating its effect to changes in deposition rate and coating topography.

Figure 37A shows the schematic layout of our SEM setup during electroplating. An Au thin-film electrode on an electron transparent silicon nitride membrane serves as the working electrode. Copper was electroplated from an acidic copper sulfate solution and imaged in situ (Figure 37B). This resolution approaches that of Schneider et al. (2014) despite the low-resolution tungsten filament SEM used. Higher resolutions will be achieved with FEG SEM in future experiments. After the deposition, the morphology of the flushed and dried Cu films was inspected in SEM (Figures 37C and D).

Experiments performed outside the SEM obtained a very high correlation between deposit thickness and electrical charge, as indicated by the results in Figure 38. From integration of the current, we calculated that

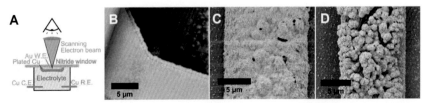

**Figure 37** The electrochemical cell (A) constitutes a working electrode of Au behind an electron transparent membrane of silicon nitride. Two pure Cu wires serve as the counter electrode and the pseudo-reference electrode. (B) In situ SEM allows time-resolved imaging of the growth. (C, D) Analysis of the electrodes after disassembly of the cell reveals different growth morphologies depending on the deposition conditions imposed on the electrode.

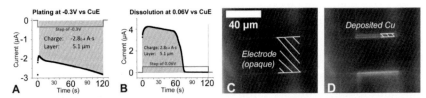

**Figure 38** Electroplating was demonstrated by optical imaging on a gold electrode (A). Layer thickness was calculated based on the total charge during the deposition. During dissolution (B), the current dropped to zero as the copper became completely dissolved. The discrepancy in copper dissolved versus copper plated occurred with the second decimal, which indicates that potentially unwanted side reactions are not a concern. (C) and (D) show the in situ optical microscopy growth of copper on an electrode.

the layer of deposited copper had a thickness of 5.1 $\mu$m (Figure 38A). During dissolution (Figure 38B), we measured the same charge before all the deposited material was consumed and the electrode became passivated.

In summary, we have developed an electrochemical cell for in situ SEM to serve as a flexible alternative to TEM when large liquid volumes are advantageous. We were able to correlate the charge to the measured film thickness and study the copper growth on the Au electrode.

## REFERENCES

Hyde, M.E., & Compton, R.G. (2003). A review of the analysis of multiple nucleation with diffusion controlled growth. *Journal of Electroanalytical Chemistry, 549,* 1–12. doi:10.1016/S0022-0728(03)00250–X.

Jensen, E., Købler, C., Jensen, P.S., & Mølhave, K. (2013). In situ SEM microchip setup for electrochemical experiments with water-based solutions. *Ultramicroscopy, 129,* 63–69. doi:10.1016/j.ultramic.2013.03.002.

Radisic, A., Vereecken, P.M., Hannon, J.B., Searson, P.C., & Ross, F.M. (2006a) Quantifying electrochemical nucleation and growth of nanoscale clusters using real-time kinetic data. *Nano Letters, 6,* 238–242. doi:10.1021/nl052175i.

Radisic, A., Vereecken, P.M., Searson, P.C., & Ross, F.M. (2006b) The morphology and nucleation kinetics of copper islands during electrodeposition. *Surface Science, 600,* 1817–1826. doi:10.1016/j.susc.2006.02.025.

Schneider, N.M., Hun Park, J., Grogan, J.M., Kodambaka, S., Steingart, D.A., Ross, F.M., & Bau, H.H. (2014). Visualization of active and passive control of morphology during electrodeposition. *Microscopy and Microanalysis, 20,* 1530–1531. doi:10.1017/S1431927614009386.

# Integrated Correlative Light and Electron Microscopy (iCLEM) with Confocal Optical Sectioning of Cells Under Vacuum and Near-Native Conditions to Investigate Membrane Receptors

**J. Sueters\*, N. Liv, P. Kruit, J.P. Hoogenboom**
Delft University of Technology, Faculty of Applied Sciences, Department of Imaging Physics, Delft, the Netherlands
*Corresponding author: e-mail address: J.Sueters@tudelft.nl

Understanding the role of biomolecules in the regulation of cellular processes (i.e., disease development by disfunction of membrane receptors) requires their observation under near-native, live-cell conditions. Fluorescence microscopy (FM) is by far the predominant imaging technique to achieve this. Although FM allows identification of biomolecules, its resolution is limited by diffraction to about 200 nm, or 20–50 nm for super resolution techniques. Electron microscopy (EM) is capable of improving the resolution to sub-nanometer levels, which is required to study biomolecules and their structural environment. Combining both techniques in an integrated correlative light- and electron microscopy (iCLEM) system (Zonnevylle, 2013), makes fast biomolecule identification within the cell ultrastructure possible at high resolution and sensitivity (Liv, 2013), as shown in Figures 39A and B. However, correlating the live-cell FM image with images of fixed cells in EM is rather complex, and in some cases it may even be impossible.

Electron microscopy of specimens in liquid has made great progress in development and applicability over the last decade. Samples under near-native cell conditions are protected from a vacuum in the EM chamber by means of a liquid cell holder. This special holder contains SiN membranes or glass substrates that are transparent for electrons or photons, respectively. We recently managed to develop a holder that enables simultaneous widefield fluorescence (WF) and liquid scanning EM (SEM) on the same region of interest in an iCLEM system (Figures 40A and B). This allows EM investigation of regions of interest of cells under near-native conditions

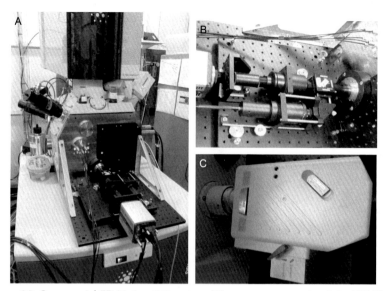

**Figure 39** Commercial FEI Verios 460 system (A), nonvacuum parts of the WF iCLEM platform (B), and commercial Nikon C2 module (C).

that are first identified and localized with WF, with limited EM resolution loss (Liv, 2014). However, SEM only investigates the upper area of the holder, whereas WF collects fluorescence from different sample depths in the imaging area. Hence, a direct correlation between WF and liquid EM image is relatively inaccurate due to differences in $z$-resolution.

An improved correlation in iCLEM requires the optical sectioning capabilities of confocal microscopy (CM). For this reason, we are currently working on an integrated setup that combines SEM with CM by means of a commercial Nikon C2 module (Figure 39C). We first plan to demonstrate this setup by studying membrane proteins on the cell surface. The $z$-resolution of 0.8 $\mu$m achievable with CM will allow localization and functional studies of membrane receptors in this iCLEM system. Further development of integrated microscopes like this one will allow CLEM to become a powerful tool for industrial applications, fundamental biological research, and medical diagnostics.

## ACKNOWLEDGMENTS

This work is in collaboration with Delmic BV. We want to thank Ruud van Tol and Carel Heerkens for their assistance, and STW for financial support.

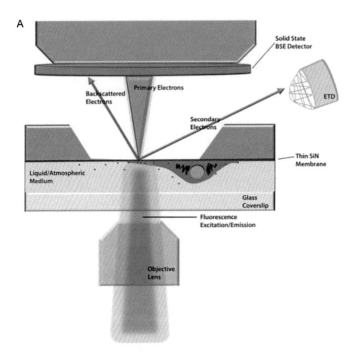

**Figure 40** Overview of CLEM system with cell holder. (A) Schematic illustration of the simultaneous observation with fluorescence and scanning electron microscopy of a sample shielded from the vacuum by a thin, electron-transparent membrane. (B) Pictures of CLEM holder (Liv et al., 2014).

# REFERENCES

Liv, N., Zonnevylle, A.C., Narvaez, A.C., Effting, A.P.J., Voorneveld, P.W., Lucas, M.S., et al. (2013). Simultaneous correlative scanning electron and high-NA fluorescence microscopy. *PLoS ONE, 8*(2), e55707.

Liv, N., Lazic, I., Kruit, P., & Hoogenboom, J.P. (2014). Scanning electron microscopy of individual nanoparticle bio-markers in liquid. *Ultramicroscopy, 143,* 93–99.

Zonnevylle, A.C., van Tol, R.F.C., Liv, N., Narvaez, A.C., Effting, A.P.J., Kruit, P., & Hoogenboom, J.P. (2013). Integration of a high-NA light microscope in a scanning electron microscope. *Journal of Microscopy, 252,* 58–70.

# In Situ Dynamic ESEM Observations of Basic Groups of Parasites

Š. Mašová[a,b], E. Tihlaříková[a], V. Neděla[a,*]

[a]ASCR, Institute of Scientific Instruments, Královopolská 147, 612 64 Brno, Czech Republic
[b]Department of Botany and Zoology, Faculty of Science, Masaryk University, Kotlářská 2, 611 37 Brno, Czech Republic
*Corresponding author: e-mail address: vilem@isibrno.cz

## INTRODUCTION

Helminth parasite infections may cause major diseases in humans and animals, so they are prime targets of veterinary and medical research. For morphological studies of parasites, as a part of taxonomy, SEM is commonly used. Sometimes only a small number or only one sample of a rare parasite is available and cannot be used in conventional SEM because the sample has to be fixed, dehydrated, and coated before it can be observed. SEM condemns samples to be destroyed so that further analysis is not possible (e.g., for molecular study or depositing as type material in a museum). Environmental scanning electron microscopy (ESEM) based on new methods (Neděla 2010; Neděla et al., 2015) and instrumentation (Jirák et al., 2010) can be used for the advanced study of parasites in their native state. It was shown by Tihlaříková, Neděla, and Shiojiri (2013) in a study of surviving mites. Only a small number of parasites were investigated using ESEM, having benefits in the unnecessary preparation to conventional SEM, which considerably reduces their post–SEM value as described by Buffington, Burks, & McNeil (2005) in insects.

The aim of this work is to show new results of two groups of helminth parasites (Nematoda, *Contracaecum osculatum;* and Acanthocephala, *Corynosoma pseudohamanni*) with minimal shape and volume distortions (keeping them acceptable for taxonomical research) and to emphasize the advantages of dynamical in situ experiments in this field of science.

## SAMPLE PREPARATION AND OBSERVING CONDITIONS

The most suitable treatment and optimal environmental conditions for observing selected parasites were found. Already-fixed samples (70% ethanol or 4% formaldehyde solution) were washed and rinsed with a drop of distilled water. Observations were made using the experimental ESEM

AQUASEM II (Neděla, 2010) equipped with an ionization detector of secondary electrons, a specially designed hydration system and Peltier cooled specimen holder. A unique method for reaching and keeping thermodynamic conditions was used (Tihlaříková, Neděla, & Shiojiri, 2013). The samples were cooled to 1 °C and observed in a high-pressure water vapor environment of 690–670 Pa, probe current of 90 pA, and beam accelerating voltage of 20 kV. Samples were placed on a Peltier cooled silicone holder into a drop of water. Consequently, the water was slowly evaporated from the sample (see Figures 41 and 42). Using our ESEM, it is possible to keep and observe susceptible wet samples for a longer time in low-beam-current conditions.

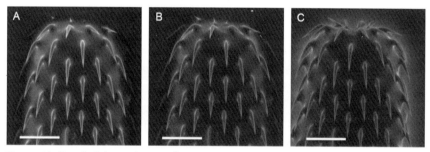

**Figure 41** Detail of the proboscis of *Corynosoma pseudohamanni* and its sequential drying documented by ESEM AQUSEM II. Scale bar: 100 μm. Observation parameters were as follows: cooling temperature 1 °C; pressure of water vapor 680 Pa; distance between the sample surface and the second pressure-limiting aperture 2.7 mm; accelerating voltage 20 kV; and probe current 90 pA.

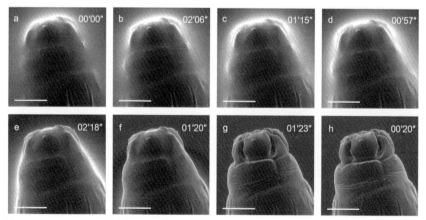

**Figure 42** Sequential drying of *Contracaecum osculatum* documented by ESEM AQUSEM II. Scale bar: 100 μm. Observation parameters were as follows: cooling temperature 1 °C; pressure of water vapor 690–670 Pa; distance between the sample surface and the second pressure-limiting aperture 2.7 mm; accelerating voltage 20 kV; and probe current 90 pA.

## RESULTS AND CONCLUSIONS

In this study, we show that ESEM allows the examination of specimens in a fully hydrated state and with minimal previous treatment. The surface was imaged in a wet state without the covering of a thick water layer to reveal microstructural specifics of samples. In situ observation of nonconductive parasitological samples was free of charging artefacts, and the visibility of topographical structures was sufficient for taxonomical research with no occurrence of artefacts and shape deformities. Eight time-lapse measurements of four lengths (1—lower end of papilla to start of ridge on body; 2—from one edge of the interlabium to the second; 3—height; and 4—length of head papilla; see Figure 43) under stable pressure conditions and low local heating by a low-probe-current electron beam were performed. Negligible changes on nematode parasite tissues were proven by measurement (Figure 43). Objects are not deformed and remain in almost the same state.

This method can be used effectively in morphological and morphometrical studies on parasites where valuable and unique specimens sometimes exist in a small number. A very slow and well controlled decrease in humidity allows the visualization of the surface topography of the sample to be free of distortions and changes in shape or dimension. Samples are well preserved from a taxonomical point of view as well. The optimal state of the sample is shown in Figure 42G. Highly visible morphology is shown in Figure 42H, but to the detriment of morphometrical data, as shown in Figure 42H. Both Figures 42G and h show a lateral view of the well-retained subventral lip equipped with one elliptical single papilla and amphid. Interlabia with undivided tip are also highly visible, as is the cuticular striation behind the lips.

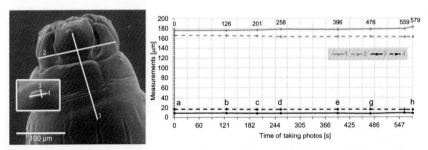

**Figure 43** Time-lapse measurements of four selected dimensions of *Contracaecum osculatum* from Figure 43, allowing the determination of shape changes in the process of specimen study in ESEM.

The suitability of ESEM for the study of parasites in a wet state and with minimal preparation was proved by our results. We introduce advanced ESEM methods continuing from the previous work by Lopes Torres *et al.* (2013), Maia-Brigagão and de Souza (2012), and Mašová *et al.* (2010). This method also will be applied with other types of parasites, and future research will be using ESEM with live samples of parasites.

## ACKNOWLEDGMENTS

This work was supported by the Czech Science Foundation: grants GA14-22777S and P505/12/G112, and the Ministry of Education, Youth, and Sports of the Czech Republic (LO1212), together with the European Commission (ALISI No. CZ.1.05/2.1.00/01.0017). The authors are grateful to the staff of the Antarctic Expedition 2014 in the Czech Antarctic Station, and J. G. Mendel on James Ross Island for providing acanthocephalan specimens.

## REFERENCES

Buffington, M.L., Burks, M. L., & McNeil, L. (2005). Advanced techniques for imaging parasitic *Hymenoptera* (*Insecta*). *American Entomologist, 51*, 50–56.

Jirák, J., Nedĕla, V., Černoch, P., Čudek, P., & Runštuk, J. (2010). Scintillation SE detector for variable pressure scanning electron microscopes. *Journal of Microscopy, 239, 3,* 233–238.

Lopes Torres, E.J., de Souza, W., & Miranda, K. (2013). Comparative analysis of *Trichuris muris* surface using conventional, low-vacuum, environmental, and field emission scanning electron microscopy. *Veterinary Parasitology, 196*, 409–416.

Maia-Brigagão, C, & de Souza, W. (2012). Using environmental scanning electron microscopy (ESEM) as a quantitative method to analyse the attachment of Giardia duodenalis to epithelial cells. *Micron, 43*, 494–496.

Mašová, Š., Moravec, F., Baruš, V., & Seifertová, M. (2010). Redescription, systematic status, and molecular characterisation of *Multicaecum heterotis Petter, Vassiliadès et Marchand,* 1979 (*Nematoda: Heterocheilidae*), an intestinal parasite of *Heterotis niloticus* (*Osteichthyes: Arapaimidae*) in Africa. *Folia Parasitologica, 57*, 280–288.

Nedĕla, V. (2010). Controlled dehydration of a biological sample using an alternative form of environmental SEM. *Journal of Microscopy, 237*, 7–11.

Nedĕla, V., Tihlaříková, E., & Hřib, J. (2015), The low-temperature method for the study of coniferous tissues in the environmental scanning electron microscope. *Microscopy Research Techniques, 78*, 13–21. doi: 10.1002/jemt.22439.

Tihlaříková, E., Nedĕla, V., & Shiojiri, M. (2013). In situ study of live specimens in an environmental scanning electron microscope. *Microscopy and Microanalysis, 19*, 914–918.

# Determination of Nitrogen Gas Pressure in Hollow Nanospheres Produced by Pulsed Laser Deposition in Ambient Atmosphere by Combined HAADF-STEM and Time-Resolved EELS Analysis

**Sašo Šturm***
Jožef Stefan Insitute, Jamova cesta 39, 1000 Ljubljana
*Corresponding author: e-mail address: saso.sturm@ijs.si

Structures with hollow interiors, such as hollow nanospheres, have recently received considerable scientific attention owing to their unique properties, which could facilitate breakthrough applications in various fields of nanoscience. Physical processing routes like pulsed–laser ablation (PLA) in the presence of a background gas have tremendous potential for applications because they offer flexibility in terms of the choice of materials to be ablated and, at the same time, the ability to produce well-defined, hollow nanospheres. It was shown previously that complex hollow nanospheres filled with nitrogen gas can be produced via a single-step procedure by the ablation of a metallic or alloy-based target into ambient nitrogen gas (Šturm et al., 2010, 2013).

The spatially resolved electron energy loss spectroscopy (EELS) analyses performed in the void and in the shell region of the hollow nanospheres consistently showed the presence of nitrogen only in the voids (Figure 44A). Figure 44B shows the resulting N–K edge for Co-Pt and Fe-(SmTa) systems, which is presented together with the reference standard spectrum of $N_2$ gas obtained from air. The fine structure for all these N–K edges is distinctive for molecular nitrogen and is characterized by a sharply peaked edge at 401 eV, followed by a broad continuum. Following these results, the nanospheres observed in HAADF-STEM images (Figure 44A) were defined as hollow spheres filled with nitrogen gas.

To unambiguously determine if measured N signal corresponds to the presence of nitrogen gas trapped inside the hollow spheres, time-resolved EELS measurements combined with the HAADF-STEM imaging were performed. EEL spectra were acquired as a function of time until the perforation of the wall of the sphere, which was achieved by the intense electron probe (Figure 45A). The corresponding background subtracted

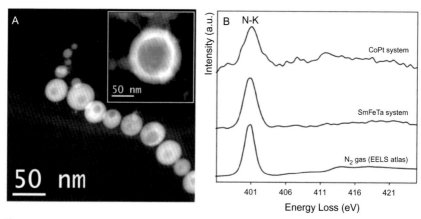

**Figure 44** (A) HAADF-STEM image of hollow spheres. The representative hollow sphere is shown in the inset. (B) Vertically displaced background-subtracted N-K ionization edges obtained from the Co-Pt and Fe-(SmTa) systems and air are distinctive for molecular nitrogen.

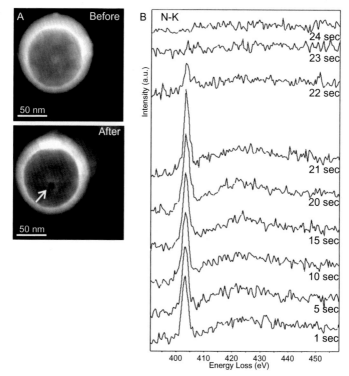

**Figure 45** (A) HAADF-STEM images of a hollow sphere before and after the experiment. The pinhole created by the intense electron probe is marked by an arrow. (B) The corresponding vertically displaced, time-resolved EEL spectra of the N-K edge with the indicated exposure time.

N-K edges measured as a function of the exposure time are shown in Figure 45B. The N-K edge signal remains nearly constant in the first 20 s and drops abruptly to the background noise in the time frame of about 1 s, which clearly demonstrates the release of nitrogen gas from the interior of the sphere through the pinhole created by the intense electron probe.

Quantitative analysis of the nitrogen density and pressure in Al, Fe-(SmTa), and Co-Pt systems was additionally performed. The number density of nitrogen atoms ($n$) in the voids of the nanospheres was determined by the following equation [2]: $n = I_N/(\sigma_N I_{ZL} d)$, where $I_N$ and $I_{ZL}$ are the intensities measured from the N-K ionization edge and the zero-loss peak, respectively; $\sigma_N$ is the angle-integrated hydrogenic cross section for the nitrogen K-shell ionization, calculated for the experimental collection angles; and $d$ represents the measured diameter of the void. To calculate the pressure in the void accurately, with respect to the given density and the temperature range, a standard correction of an ideal gas law using a virial expansion was applied, as follows: $P = nkT(1 + nB/N_A + n^2 C/N_A^2 + \ldots)$, where $k$ represents the Boltzmann constant; T is the absolute temperature,4; $N_A$ stands for the Avogadro constant; and B and C are the second and third virial coefficients, respectively. The calculated pressures ranged between 25 bars for the Co-Pt system up to 450 bars in Fe-(SmTa) and Al systems, respectively. The total error of the calculated pressures amounts to 17%. The obtained results support the idea that gas-filled hollow spheres could be fabricated in various complex metallic systems by applying PLA in the presence of a background gas, taking into consideration that in relation to the background gas high-solubility differences between the melt and corresponding solids are achieved.

## REFERENCES

Šturm, S., Žužek Rožman, K., Markoli, B., Spyropoulos Antonakakis, N., Sarantopoulou, E., Kollia, Z., et al. (2010). Formation of core-shell and hollow nanospheres through the nanoscale melt-solidification effect in the Sm-Fe(Ta)-N system. *Nanotechnology, 21,* 485603-1–485603-8.

Šturm, S., Žužek Rožman, K., Markoli, B., Spyropoulos Antonakakis, N., Sarantopoulou, E., Kollia, Z., et al. (2013). Pulsed-laser fabrication of gas-filled hollow CoPt nanospheres. *Acta Materialia, 61,* 7924–7930.

# Platelet Granule Secretion: A (Cryo)-Correlative Light and Electron Microscopy Study

K. Engbers-Moscicka[a], C. Seinen[b], W.J.C. Geerts[a], H.F.G. Heijnen[b,c,]*

[a]Department of Biomolecular Electron Microscopy, Bijvoet Center, Utrecht University, Utrecht, The Netherlands
[b]Laboratory of Clinical Chemistry and Hematology, University Medical Center Utrecht, Utrecht, The Netherlands
[c]Cell Microscopy Core, Department of Cell Biology, University Medical Center Utrecht, Utrecht, The Netherlands
*Corresponding author: e-mail address: h.f.g.heijnen@umcutrecht.nl

Blood platelets are anucleate cells that play a central role in the arrest of bleeding after a blood vessel is damaged. Platelets circulate at variable velocities in the bloodstream, and are activated following binding to subendothelial components [von Willebrand factor (vWF), collagen] that are exposed during vascular injury. Stable adhesion to collagen promotes the release of soluble components from platelet secretory granules, leading to the formation of a thrombus. Platelets also play a crucial role in the inflammatory response (Boilard et al., 2010; Semple et al., 2011), through mechanisms that are less well characterized.

Platelet secretory alpha-granules contain adhesive proteins (vWF, fibrinogen), pro- and anti-inflammatory mediators (PF4, beta-TG), and luminal membranes harboring cytosolic components (IL1β). Timed release of granule content and cell surface expression of integral membrane components are important determinants for platelet contribution in modulating haemostatic and inflammatory responses. Platelet secretory granules are heterogeneous in morphology and content, and this heterogeneity may give rise to differential release patterns. The molecular mechanism that regulates the fine-tuning of alpha granule cargo release is not known.

The relative small thickness of adherent platelets (less than 500 nm) makes them ideally suited for whole cell electron tomography analysis. Here, we describe two CLEM methodologies to study platelet adhesion dynamics and cargo release at the ultrastructural level. Cryo-electron microscopy (cryo-EM) allows for the visualization of cells in a close-to-native state, with nanometer-scale resolution (Faas et al., 2013). Due to low contrast and low signal-to-noise ratio in images of frozen hydrated samples, it often appears difficult to localize structures of interest within heterogeneous samples. We have used iCorr-integrated correlative microscopy (Agronskaia *et al.*, 2008) with high-resolution (cryo) electron tomography to study the secretory behavior of human platelet alpha granules. iCorr is designed to automate and accelerate CLEM experiments, resulting in fluorescent labeling information combined with 3D ultrastructural information.

Platelets, isolated from freshly drawn human whole blood were allowed to settle on fibrinogen coated copper grids, and stimulated with collagen related peptide. Flow devices specially developed for simultaneous light microscopy imaging and electron microscopy transfer were used to follow membrane dynamics and granule release patterns under physiological flow. Release of vWF was detected using combined immunolabeling with a fluorophore and protein A gold.

Our results show that we are able to correlate fluorescent signals, arising from released vWF, to the apical cell surface of the adherent platelets (Figures 46–47). From dual-axis tomography analysis on flow-adherent cells, it appeared that subsets of vWF-containing granules migrate in a downstream fashion, while others remain stationary in the cell center. Downstream-oriented granules release vWF as strings in the direction of the flow (Figures 46A,B). Similar results were obtained when adherent platelets were vitrified in liquid ethane, followed by correlative analysis by the EM-integrated iCorr workflow (FEI Company) (Figure 47). Low-dose tomographic analysis revealed subtle substructural modifications at the limiting alpha granule membrane that had not been observed before (Figure 47D). Whether these subtle membrane changes, which were visualized only after vitrification, give us a clue how platelets manage to differentially release their granule contents remains to be seen.

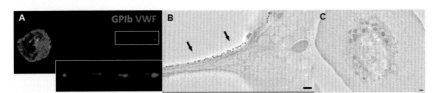

**Figure 46** Images showing flow-driven granule dynamics and polarized vWF release. (A) Confocal image showing vWF (red fluorescence) in alpha granule subsets migrated in a downstream fashion (arrowhead), while others remain stationary in the cell center. Inset: Release of vWF strings. GPIb (green), platelet receptor for vWF. (B) Dual axis tomographic slice of whole adherent platelet. vWF strings (immuno-gold particles indicated by the black arrows) are released in the direction of the flow. (C) Tomographic slice from an adherent platelet showing immuno-gold labeling of vWF on the surface of the platelet cell body. Scale bar: 100 nm.

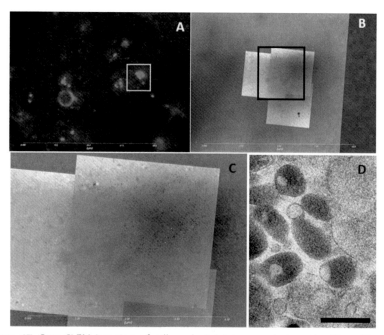

**Figure 47** Cryo-CLEM imaging of adhered platelets recorded on the iCorr. Platelets were double-labeled for vWF with a fluorophore (A) and protein A gold (B,C), and then plunge frozen in liquid ethane. Cryo-TEM images (B) superimposed on the fluorescent image (white box in panel A). Image C zooms in on the specific immunogold labeling for vWF of the area box in B. (D) Slice of a cryo-tomogram revealing substructural details of the limiting alpha granule membrane that had not been observed before. Scale bar: 150 nm.

In conclusion, cryo–CLEM and cryo–tomography are powerful techniques that enable fast immobilization and capturing of membrane dynamics that occur during platelet activation, such as fusion pore formation and localized and timed release phenomena.

## REFERENCES

Agronskaia, A.V., Valentijn, J.A., van Driel, L.F., Schneijdenberg, C.T.W.M., Humbel, B.M., van Bergen en Henegouwen, P.M.P., et al. (2008). Integrated fluorescence and transmission electron microscopy. *Journal of Structural Biology, 164,* 183–189.

Boilard, E., Nigrovic, P.A., Larabee, J., Watts, G.F.M., Coblyn, J.S., Weinblatt,M.E., et al. (2010). Platelets amplify inflammation in arthritis via collagen–dependent microparticle production. *Science, 327,* 580–583.

Faas, F.G.A., Bárcena, M., Agronskaia, A.V., Gerritsen, H.C., Moscicka, K.B., Diebolder, C.A., et al. (2013). Localization of fluorescently labeled structures in frozen–hydrated samples using integrated light electron microscopy. *Journal of Structural Biology, 181,* 283–290.

Semple, J.W., Italiano, J.E., Jr., & Freedman, J. (2011). Platelets and the immune continuum. *Nature Reviews Immunology, 11,* 264–274.

## ACKNOWLEDGMENTS

Members of the scientific organizing committee:

- Kristian Mølhave, DTU Nanotech—Department of Micro and Nanotechnology, University of Denmark, 2800 Kgs. Lyngby, Denmark
- Niels de Jonge, INM

Christine Hartmann organized the logistics of the conference and edited the abstract booklet. The INM provided financial support through Eduard Arzt.

The following sponsors are greatly acknowledged:

- CEOS GmbH, Heidelberg, Germany (high-level sponsor)
- European Microscopy Society
- Dens Solutions, Delft, the Netherlands (high-level sponsor)
- Deutsche Gesellschaft fuer Elektronenmikroskopie
- FEI, Hillsboro, Oregon
- E.A. Fischione Instruments, Inc., Export, Pennsylvania
- Gatan Inc., Pleasanton, California
- JEOL Ltd., Tokyo 196-8558, Japan (high-level sponsor)
- Protochips, Inc., Raleigh, North Carolina (high-level sponsor)

Photograph of participants CISCEM 2014.

CHAPTER TWO

# Progress and Development of Direct Detectors for Electron Cryomicroscopy

## A.R. Faruqi[1], Richard Henderson, Greg McMullan

MRC Laboratory of Molecular Biology, Francis Crick Ave., Cambridge Biomedical Campus, Cambridge CB2 0QH, UK
[1]Corresponding author: e-mail address: arf@mrc-lmb.cam.ac.uk

## Contents

*Advances in Imaging and Electron Physics*, Volume 190
ISSN 1076-5670
http://dx.doi.org/10.1016/bs.aiep.2015.03.002

103

# 1. INTRODUCTION

The main application of direct electron detectors discussed in this chapter is in single-particle electron microscopy (SPEM) or cryomicroscopy or cryo-EM of frozen hydrated biological specimens, a technique used to obtain structures of biological macromolecules to high (i.e., near-atomic) resolution from large numbers of images of individual molecules without the need for crystals. Electron cryomicroscopy (cryo-EM) in its current form emerged from Jacques Dubochet's development of a method for rapid freezing of single particles in thin films of amorphous ice (Adrian, Dubochet, Lepault, & McDowall, 1984; Dubochet et al., 1988). Single-particle image processing procedures were first applied to negatively stained specimens before being refined for use on images of frozen hydrated specimens (Frank, 2009). Before the advent of direct electron detectors, cryo-EM had been used, mainly with photographic film, to obtain high-resolution structures of relatively large macromolecular assemblies such as viruses. The newly developed direct detectors, together with the associated new methods of data acquisition, produce higher-resolution structures and require fewer single particles. They can also be used to analyze the structures of a range of smaller-molecular-weight molecules that were previously beyond the reach of cryo-EM.

SPEM is a technique used to obtain images from individual molecules frozen in a thin layer of vitreous ice, with their structure being close to native. Images of individual molecules have poor contrast due to the weak scattering by the constituent carbon, nitrogen, oxygen and hydrogen atoms, and poor signal-to-noise ratio due to the limited electron dose permitted before radiation damage builds up. Extracting high-resolution information requires the aligning and averaging of tens of thousands (or more) of individual images. Molecules with high symmetry require fewer images and images of molecules with higher molecular weight have higher-contrast images and are easier to analyze (Grigorieff & Harrison, 2011).

In recent years, a number of outstanding near-atomic-resolution structures of macromolecular complexes have been obtained using single-particle cryo-EM, reviewed briefly in several studies (Faruqi, Henderson, & McMullan, 2013; Kühlbrandt, 2014; Henderson, 2015) and covered more fully in other publications (Agard, Cheng, Glaeser, & Subramaniam, 2014; Liao, Cao, Julius, & Cheng, 2014; Bai, McMullan, & Scheres, 2015). Kühlbrandt (2014) has commented that such unprecedented resolution, though expected on theoretical grounds (Henderson, 1995), still came as a great and pleasant surprise. While better software and increased computing power have contributed to this success, the introduction of direct electron detectors is clearly the most important factor. On the same basis, further improvements are expected once detectors with even better performance become available.

A comparison of the performance of three currently available commercial direct detectors was published recently (McMullan, Faruqi, Clare, & Henderson, 2014). Electronic detectors based on phosphor coupled charge-coupled devices (CCDs) have been used for cryo-EM for the past 20 years, particularly in two-dimensional (2D) electron crystallography (Faruqi & Subramaniam, 2000; Sander, Golas, & Stark, 2005; Bammes, Rochat, Jakana, & Chiu, 2011). It is expected that direct detectors with a higher detective quantum efficiency (DQE) will also lead to improved data for electron crystallography and tomography (Lucic, Forster, & Baumeister, 2005; Szwedziak, Wang, Bharat, Tsim, & Löwe, 2014), as well as for single-particle cryo-EM. Recording electron diffraction data from small three-dimensional (3D) microcrystals has also been made possible by direct detectors based on hybrid detector technology—namely, Medipix2 (Campbell, 2011; Nederlof, van Genderen, Li, & Abrahams, 2013). Since this chapter is focused mainly on biological cryo-EM, many other applications outside this field, such as those in materials science and condensed matter physics are not included, though many of these applications have also benefited from improved detectors. How detectors are evaluated in terms of size, efficiency, resolution, and other factors is described in the section entitled "Detector Basics," later in this chapter; such measurements are very useful for making an objective direct comparison between different detectors (McMullan, Chen, Henderson, & Faruqi, 2009; Ruskin, Yu, & Grigorieff, 2013; McMullan et al., 2014).

Several publications and reviews have covered in some detail the desirable (and in some cases essential) properties required in a detector for cryo-EM (Faruqi et al., 2005a; Faruqi, 2007; McMullan, Chen, Henderson, & Faruqi, 2009; Faruqi & McMullan, 2011), so this chapter will merely summarize the most important points for this review. As the discussion is dealing with specimens that are damaged easily by radiation, the most important requirement is a high DQE, which incorporates detection efficiency, spatial resolution and noise properties of the detector in its definition. The DQE of a detector is given by the ratio of the square of signal to noise at output compared with input:

$$DQE = (S/N)_{out}^{2}/(S/N)_{in}^{2},$$

where S is the signal and N the noise. It is important to note that, because the detector always adds some noise to the signal, the DQE is always less than 1. When used for imaging, the DQE needs to include the effects of spatial resolution and when given as a function of spatial frequency, $\omega$, it is given by DQE($\omega$) (Dainty & Shaw, 1974; Meyer & Kirkland, 2000; McMullan, Chen, Henderson, & Faruqi, 2009), where

$$DQE(\omega) = DQE(0)^*MTF^2(\omega)/NPS(\omega),$$

with MTF($\omega$) being the modulation transfer function as a function of spatial frequency and NPS($\omega$) being the normalized noise power spectrum, also as a function of spatial frequency. For pixelated detectors, the important concept of the Nyquist frequency, given by the inverse of twice the pixel size, is used frequently in the evaluation of individual detector properties. The Nyquist limit is also useful in comparing different detector properties at this spatial frequency, which is independent of the pixel size in different detectors (McMullan, Chen, Henderson, & Faruqi, 2009; Faruqi & McMullan, 2011; Ruskin et al., 2013; McMullan et al., 2014).

A review of high-resolution structures obtained with cryo-EM published several years ago (Grigorieff & Harrison, 2011), before the advent of the direct electron detectors, listed nine different near-atomic-resolution structures of icosahedral viruses. The single particles chosen for that review were all quite large, in the range of 17–150 MDa, some with high symmetry, which makes their orientation determination much easier due to the high-contrast images they produce and consequent ease of averaging data from a number of particles. Apart from one structure, all the research cited used photographic film, which was previously accepted to be the best recording medium with the highest DQE (Sander et al., 2005; McMullan, Chen, Henderson, & Faruqi, 2009). A similar review today, describing the most recent high-resolution structures, would include a majority based on the use of direct electron detectors, usually with back-thinned complementary metal oxide semiconductor (CMOS) technology, as discussed in the "Detector Basics" section [see also Bai et al. (2015)]. There are a number of reasons for this change, and this chapter aims to review some of the properties that make direct electron detectors so attractive for cryo-EM.

## 2. DETECTORS: INDIRECT AND DIRECT

The terminology of indirect and direct detectors predates the work on detectors for cryo-EM by more than a decade. Initially used for X-ray detection, *direct detection* referred to X-ray photons detected in gas-based detectors, which converted the energy of the incident photon into electron-positive ion pairs and recorded counts after amplification within the gas in multiwire proportional chambers (Faruqi, 1991). On the other hand, *indirect detection* referred to detectors in which the incoming photons were absorbed in a phosphor or scintillator; the emitted light was then detected and recorded by a vidicon camera or a CCD (Arndt, 1986). Following the application of phosphor–fiber optics coupled CCD detectors to X-ray diffraction experiments, detectors based broadly on similar principles were also developed for use in cryo-EM (for a review of earlier work on CCD-based cryo-EM, see Faruqi & Subramaniam, 2000; Sander et al., 2005; Clare & Orlova, 2010), but the DQE at high spatial frequencies was found to be worse than film at 300 keV; a comparison of the relative DQE values for film- and CCD-based detectors (and for Medipix2 at 120 keV) can be found in McMullan, Chen, Henderson, and Faruqi (2009) and Faruqi and McMullan (2011). An understanding of the physical reasons behind this, such as light scattering in the phosphor and fiber optics, which were thought to be the cause of lower DQE in CCD-based indirect detectors (Meyer & Kirkland, 1998) was the motivation for investigating direct detectors, which avoid energy conversion to light prior to detection (Faruqi & Andrews, 1997).

Research and development on direct detection for cryo-EM started about 15 years ago in the Medical Research Council Laboratory of Molecular Biology (MRC-LMB) in Cambridge, England (Faruqi, 2001; Faruqi, Cattermole, Henderson, Mikulec, & Raeburn, 2003; Faruqi, Henderson, & Tlustos, 2005), but they were based on completely different design principles due to the nature of the radiation (namely, electrons) being detected (rather than photons). Instead of gas, the primary detection medium for electrons was a specially doped semiconductor material (usually silicon). In terms of signal formation, the signal in direct detectors is large since, on average, one electron–hole pair is created per 3.6 eV energy deposited in silicon by the incoming electron, which results in a high signal-to-noise ratio due to the large number of electron–hole pairs created by every incident electron. Early attempts at using direct detectors were focused on Medipix2 (Llopart & Campbell, 2003), which is a hybrid pixel detector in which the silicon electron sensor is separated from the electron counting layer (Llopart, Campbell, Dinapoli, San Secundo, & Pernigotti, 2002). They are discussed briefly in the section entitled "Imaging Problems and Solutions Provided by Direct Detectors," later in this chapter (also see Faruqi et al., 2003; Faruqi et al., 2005b; McMullan et al., 2007) and have been reviewed previously in this journal (Faruqi, 2007). Two applications of Medipix2 in cryo-EM are discussed in the "Imaging Problems and Solutions Provided by Direct Detectors," section. However, despite some very clear advantages at lower energies (<120 keV) over indirect detectors, such as phosphor fiber-optic CCDs and film, they were found to have significant shortcomings for electron energies of primary interest for cryo-EM; namely 200–300 keV (McMullan, Chen, Henderson, & Faruqi, 2009). It will become clearer after discussions on sensor back-thinning (e.g., McMullan et al., 2009) why it would be extremely difficult to build a sufficiently thin hybrid detector with an adequately high DQE for 300-keV operation.

Preliminary trials with monolithic active pixel sensors (MAPS), designed in CMOS technology originally for optical applications, gave encouraging results for electrons with 40- and 120-keV energy (Xuong et al., 2004; Faruqi et al., 2005a). However, they also highlighted several design features in the original chip, which had to be improved considerably to achieve a successful detector, as discussed in the section "Detector Basics," later in this chapter (also see Faruqi et al., 2005). Very encouraging results were also obtained at electron energies up to 400 keV (Jin et al., 2008). The improvements that made the most impact, some of which have been discussed previously (Faruqi & McMullan, 2011), were (1) the ability to address larger areas with increased numbers of pixels; (2) radiation hardening, allowing long-term use in the electron microscope without the performance being degraded; (3) back-thinning the sensor, making it possible to obtain high DQE at all spatial frequencies; and (4) fast continuous readout, allowing movie-mode recording.

Many groups have used this fourth key advantage offered by direct detectors; namely, the ability to provide dose fractionation by recording a time series of images (movies) from the same specimen (Brilot et al., 2012; Campbell et al., 2012). In parallel with the development of new detectors, new types of computer programs for image processing have been developed. One that is especially notable because of its successful introduction of maximum likelihood methods is a software package known as RELION (Scheres, 2012). RELION uses statistical methods to deal with heterogeneous specimens. It is also able to obtain more precise estimations of the orientation of individual particles, which is essential for the calculation of improved structures (Scheres, 2012). For time-invariant structures, the individual movie frame images could be simply summed to give an integrated image, but one of the problems during imaging is that there is some beam-induced specimen movement as well as radiation damage. The new CMOS cameras are able to collect sequential images, and the computational algorithms then can treat them individually, correcting for specimen movement before integration and producing a sharper image (Bai, Fernandez, McMullan, & Scheres, 2013; Li et al., 2013; Brilot et al., 2012).

The combined benefits of high DQE, which gives better images with better signal-to-noise ratio, and fast readout have allowed the collection of dose-fractionated images in movie mode, which was not possible with indirect detectors in the past. The section entitled "High-Resolution Structures Obtained Recently with Direct Detectors," later in this chapter, gives examples of some achievements in obtaining near-atomic resolution for macromolecular structures made possible by direct detectors and newly developed software.

## 3. DETECTOR BASICS

The development of present-day direct detectors, which is relevant for cryo-EM, follows a lengthy period of experience with earlier detectors, which created a framework to understand what would make a perfect electron detector. Such a perfect detector is still some way off in the future, but this chapter deals with the story approximately from the completion of the indirect detector phase up to the present. Some of the historical part has been covered in articles and reviews written only a few years back, which covered film, indirect detectors based on CCDs, hybrid pixel detectors based on Medipix2, and the early developments on CMOS detectors (e.g., Faruqi & Subramaniam, 2000; Faruqi & McMullan, 2011). In this review we will concentrate on monolithic active pixel sensors (MAPS) based on back-thinned CMOS technology, which has proven to be the most successful in achieving this goal (McMullan, Clark, Turchetta, & Faruqi, 2009; McMullan et al., 2009; Guerrini et al., 2011b; McMullan, Turchetta, & Faruqi, 2011; Battaglia et al., 2010; Milazzo et al., 2011).

In recent cryo-EM applications, a number of non–single-particle imaging studies have also benefited from using high-DQE direct detectors. These include electron tomography, electron crystallography of 2D crystals, and electron crystallography of 3D microcrystals. In the absence of 3D crystals that are large enough to obtain data by X-ray diffraction at a synchrotron camera, it is still possible to obtain structures using electron crystallography. 3D electron crystallography has become possible on submicrometer-size crystals, without the need for access to an X-ray Free Electron Laser (XFEL) (Liu et al., 2013). In the first of these studies, Nederlof et al. (2013) used the rotation method to record diffraction patterns from lysozyme crystals on a tiled Quad-Medipix2 at 200 keV. In the second example, Nannenga, Shi, Leslie, and Gonen (2014) used a Tietz CMOS detector (http://www.tvips.com/) to record electron diffraction data using the rotation method, which was later processed with minor changes to standard crystallographic packages.

## 3.1 MAPS Based on CMOS Technology

Simplified diagrams of the cross section of a MAPS pixel are shown in Figures 1(A) and 2, along with a simplified readout scheme for a single pixel in Figure 1(B). It is worth mentioning that the three-transistor (3T) pixel design is only one of a number of different pixel designs, and other more complicated designs can be used for additional functionality and improved performance.

MAPS detectors most of the time are fabricated using a heavily p-doped silicon wafer onto which a layer of between 2 and 20 $\mu$m of lightly p-doped silicon is epitaxially grown. The epitaxial layer, known as the *epilayer*, is highly ordered and forms the sensitive layer for the detector. The epilayer is implanted with a highly p-doped layer as well as highly n-doped wells reaching into the epilayer. The n-doped wells form $N^+$ diodes, and their spacing defines the pixel size of the detectors. An incident high-energy electron passing through the epilayer generates low-energy electron–hole-pairs. These free carriers have long mean-free paths due to the highly ordered crystal lattice of the epilayer, but while the holes are free to diffuse into the surrounding layers and could eventually be collected by the heavily doped p-wells or silicon wafer, the low-energy electrons are kept in the epilayer by the potential resulting from the doping gradient to the more highly p-doped layers above and below (Turchetta et al., 2001). They will eventually be collected by the $N^+$ diodes.

In a 3-T MAPS detector, three NMOS transistors are fabricated in the upper p-doped layer. One transistor is used as a reset and allows a fixed initial reverse-bias voltage to be set on a $N^+$ diode. The reverse-biased $N^+$ diode acts as a capacitor and is discharged by any electron excitations in the epilayer that come close enough to be captured. The second transistor is configured as a source follower and passes the voltage on the $N^+$ diode to the third transistor, which is used to select the particular $N^+$ diode voltage to be read out.

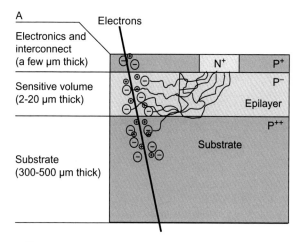

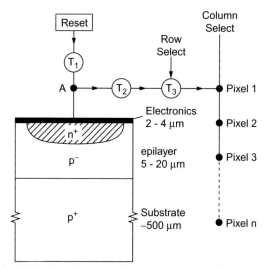

**Figure 1** (A) Typical side view of a 3T pixel in a MAPS. The incident electron leaves a trail of electron–hole pairs along the track. Electrons created within the epilayer have a sufficiently long mean free path to diffuse to the $N^+$ diode, buried in the $P^+$ well, helped by a slight potential difference due to different doping densities in $P^-$ and $N^+$ silicon. The substrate consists of highly doped silicon, which acts as an inert structure with rapid electron–hole recombination, so that it plays only a marginal role in signal generation (Turchetta et al., 2001). (B) Schematic diagram showing a simplified readout from a single pixel in a 3T CMOS detector. As shown in (A), an incoming electron results in electron–hole pairs and electrons, created mostly within the epilayer, but only electrons drift to the $n^+$ diode, forming the signal. The stray capacitance at node A is charged prior to the event by switching transistor $T_1$ on, and the electrons from the event discharge this capacitor by an amount proportional to the charge collected on the diode. Charge from the pixel is read out with the aid of a row select transistor, $T_2$, and column select transistor, $T_3$, into external analog-to-digital converters, which are used to form the image (Faruqi, 2007). (See the color plate.)

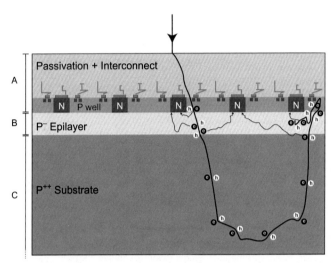

**Figure 2** Monte Carlo simulation of a single-electron trajectory through a pixel, which was backscattered in the silicon substrate. A: passivation + Interconnect; B: Epilayer; C: Substrate. The passage of the electron creates electron–hole pairs along the track, and the electrons and holes are shown as blue and red circles, respectively. Electrons created in the epilayer diffuse to the collecting diode. Toward the end of the track in the backscattered electron, there is a much higher concentration of electron–hole pairs as the electron has much lower energy, resulting in a far higher signal that is also in the wrong place (McMullan, Chen, Henderson, & Faruqi, 2009; McMullan et al., 2009). (See the color plate.)

The voltages applied to and read out from the transistors are passed to a particular pixel via metal interconnects embedded in nonconducting oxide layers fabricated on top of the p-doped layer. Finally, the whole detector is covered in a passivation layer consisting of $SiO_2$ and/or $Si_3N_4$, a few microns in thickness depending on the technology.

One of the most basic readout schemes for CMOS detectors, which was described in a previous review on direct detectors, is reproduced in Figure 1(B) (Faruqi, 2007). In the so-called rolling shutter mode, a reset voltage is applied sequentially in turn to each row of $N^+$ diodes. After reset, the change in the voltage of individual diodes reflects the amount of charge generated in the epilayer near a given diode, which is proportional to the number of incident electrons. After a fixed time, the voltages on the diodes in a given row are read out and digitized in an analog-to-digital converter located either on the chip or externally. The rolling shutter mode is intrinsically parallel and allows high readout speed, which is essential for movie-mode imaging, to be obtained. One such scheme has been implemented for a 4k × 4k sensor by Guerrini et al. (2011b). For imaging radiation-sensitive specimens in general, but especially for movie-mode imaging, it is essential to maximize the signal-to-noise ratio. Details of the various sources of readout noise are analyzed in some detail, and possible solutions to minimizing their effects are discussed in Anaxagoras et al. (2010).

## 3.2 Effects of Backscattering on Performance of CMOS Detectors

A single electron traversing the epilayer in a sensor will deposit a small amount of energy, typically resulting in the generation of about 80 electron–hole pairs per micron of silicon. The amount of energy deposited varies due to the stochastic nature of the interaction, with some events contributing much higher energies than the mean. As pointed out by McMullan et al. (2011), the behavior of 300 keV electrons gives rise to some problems. Any electron traversing the epilayer and the substrate leaves a well-defined signal in terms of amplitude and location. However, backscattered electrons from the substrate can generate a signal some distance from the point of initial impact. A simulation of a backscattering event is shown in Figure 2 (McMullan et al., 2009). Since the backscattered electrons have traveled further through silicon, they have lost energy and are thus able to deposit a larger amount of energy on their second traverse through the epilayer, producing a larger signal than the original traverse. Since the overall signal for an event is increased due to the backscattering event, the DQE at zero spatial frequency is relatively unchanged, or indeed becomes slightly higher. However, DQE at higher spatial frequencies is reduced as the backscattering contribution degrades the resolution and adds extra high-frequency noise to the event (Deptuch et al., 2007; McMullan et al., 2009).

The causes of lower DQE due to backscattering are illustrated in another Monte Carlo simulation of the behavior of 300 keV electrons in silicon (shown in Figure 3). The overall thickness of silicon chosen for the simulation, 350 $\mu$m, was similar to the thickness of a typical sensor, while the top layer (shown in dark gray in the figure) was 35 $\mu$m. While electron tracks (shown in dark gray) contribute to the "good" signal, backscattered electrons (shown in white) reduce the DQE at high spatial frequencies. The benefit of back-thinning, or removing the light gray part of the sensor, is that most of the backscattered tracks (shown in white) are eliminated. The simulation in Figure 3 demonstrates the vital importance of back-thinning in designing a high-DQE sensor, as has been verified experimentally (e.g., Deptuch et al., 2007; McMullan et al., 2009, 2014).

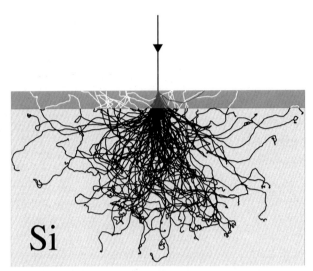

**Figure 3** Monte Carlo simulation showing trajectories of 300-keV electrons through a silicon sensor with a thickness of 350 $\mu$m with the top part (shown in dark gray) being 35 $\mu$m. The incident electron is shown in black. Electrons giving a high DQE are shown in red, apart from one electron, which is backscattered in the top layer. The white tracks show backscattered electrons that create a signal away from the impact position and decrease the DQE at high spatial frequencies (McMullan et al., 2009). This simulation suggests that a back-thinned sensor with a thickness of 35 $\mu$m should have superior DQE, and this has been confirmed (McMullan et al., 2009). (See the color plate.)

## 3.3 Radiation Damage to the Sensor During Use

Unlike CCDs in phosphor fiber–optically coupled detectors or the radiation-sensitive electronic layer in hybrid pixel detectors, CMOS detectors are directly exposed to electrons, which make them far more susceptible to radiation damage (Faruqi et al., 2005a). The radiation damage to the detector due to incident electrons at energies of interest to cryo–EM experiments (i.e., up to about 300 keV) is almost entirely caused by ionization damage, which depends on the total energy deposited by the primary electron in the sensor (Bogaerts, Dierckx, Meynants, & Uwaerts, 2003; Turchetta, 2007). The probability of displacement damage when sufficient energy is transferred to a silicon atom to knock it out of the regular lattice of silicon is negligible at these energies (Bogaerts et al., 2003). The major effect of ionization is an increase in the sensor dark or leakage current. The main consequence of an increase in dark current is a lowering of the dynamic range of the sensor with increasing doses, which ultimately prevents the sensor from being used at all. Although annealing at high temperatures or with time is known to partially reverse the effects of radiation damage, this is not a practical solution for electron microscopy. The lifetime of a CMOS sensor used in early experiments, which was designed originally for optical astronomy applications and hence was not designed with rad-hard transistors and diodes (Prydderch et al., 2003), was found to be 10–20 krads in an electron microscope; this level of rad hardness would be inadequate for normal use in electron microscopy (Faruqi et al., 2005a). The effects of increased dark current can be reduced by operating the sensor at a lower temperature, by employing a different layout for the pixel electronics, or by reducing the integration time during which both the signal and dark current are integrated. Shorter integration times are used when recording images in movie mode, making radiation damage relatively less important.

Radiation hardness in microelectronics circuits has been an important issue for numerous applications, such as reactor engineering, nuclear and particle physics, etc., where high levels of radiation are encountered. In order to reduce or eliminate the effects of radiation damage, several changes in the design of pixel electronics were suggested, including pixel layouts using enclosed layout transistors (ELTs; Bogaerts et al., 2003; Campbell et al., 2001). With improved rad-hard pixel designs, faster frame rates, and operation at −20 °C (lowering the temperature leads to reduced leakage current, in common with all silicon devices), the radiation hardness of CMOS sensors was increased by more than three orders of magnitude, which is essential for a practical detector with a useful operating lifetime of a year or more (Faruqi, Henderson, & Holmes, 2006; Battaglia et al., 2009; Guerrini et al., 2011a).

It is instructive to follow through with the design and construction of a practical detector for cryo-EM applications, and we have focused on a prototype—namely, the TEMAPS 1.0 detector (Guerrini et al., 2011a)—which was later further developed by FEI as a commercial detector named Falcon (http://www.fei.com/). To arrive at the optimal design of TEMAPS, the effects of irradiation on several parameters of interest, such as modulation transfer function (MTF), DQE, and radiation damage, were monitored. To maximize the likelihood of making the best choice, TEMAPS 1.0, with $1,280 \times 1,280$ pixels of size 14 $\mu$m, was designed with 25 different $256 \times 256$ subareas containing different pixel layouts. Detailed measurements were carried out on all the subareas, with results from two of the most

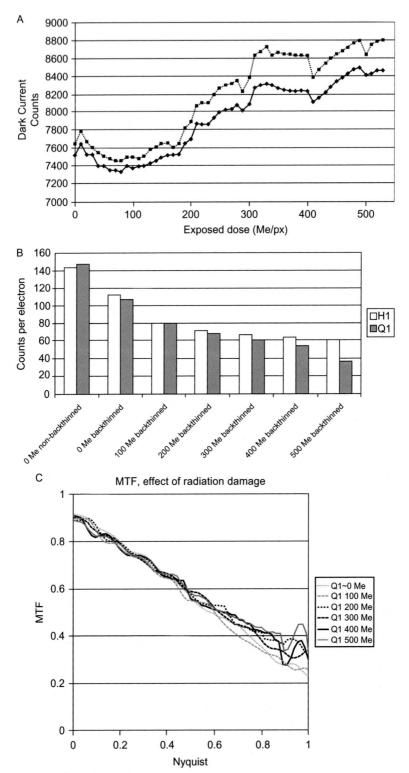

**Figure 4** See legend on opposite page.

promising shown next (Guerrini et al., 2011a) for doses extending from 0 to 500 million electrons/pixel at intervals of 10 million electrons/pixel. Figure 4(A) shows the variation in dark current in two of the chosen areas over the whole range of irradiation. Although there was an overall increase in dark current, the authors point out that, since the sensor output was digitized to 16-bit resolution, this increase simply changes the background, which could be accommodated in the measurements. The reduction in sensitivity shown in Figure 4(B) to electrons is caused by an increase in leakage current with radiation dose. The effect of radiation damage on MTF, which is quite small, is shown in Figure 4(C).

## 3.4 Practical Detectors for Cryo-EM

Due to the great potential of direct detection for electron microscopy, a number of parallel developments were initiated by different groups around the world. We have selected just three of the developments that have had the most impact, but we go into extra detail about TEMAPS (already discussed in the previous section), as we have more details about its preliminary results. TEMAPS was designed at the Rutherford Appleton Laboratory, Oxford, England, in close collaboration with MRC-LMB (Guerrini et al., 2011a), and the project was funded by a European consortium that also included FEI and Max Planck Gesselschaft. TEMAPS had 4k × 4k pixels of size 14 $\mu$m, with radiation-hardened pixel architecture that used enclosed layout transistors. In a second development of radiation hard sensors for electron microscopy, development was carried out at the Lawrence Berkeley National Laboratory (LBNL) with a detector called LDRD-RH, which was based on a chip originally developed for particle physics (Battaglia et al., 2009, 2010). A third development of rad-hard sensors was carried out by the University of California, San Diego (UCSD), in collaboration with other groups (Jin et al., 2008; Milazzo et al., 2011).

---

**Figure 4** (A) Variation of dark current as a function of irradiation in two chosen subareas as a function of dose in units of million electrons/pixel (Guerrini et al., 2011a). The blue data points come from the more rad-hard pixel architecture. (B) Variation in sensitivity in units of counts per 300-keV electron for the two chosen pixels as a function of irradiation level. The first data point at 0 Me is for a non-back-thinned sensor, whereas all the other points are for a back-thinned sensor; (Guerrini et al., 2011a). The blue data bars show data from the more rad-hard pixel architecture. (C) Plots showing the variation of MTF as a function of spatial frequency (up to Nyquist frequency) for different levels of irradiation for one chosen pixel architecture (Guerrini et al., 2011a). (See the color plate.)

A summary of the basic properties of TEMAPS are listed in Table 1. As mentioned earlier in this chapter, a commercial detector, based on the design of TEMAPS, called Falcon I (not back-thinned) or Falcon II (back-thinned), was produced by FEI (http://www.fei.com/) for use in electron microscopy; a photograph in Figure 5 shows an early version of Falcon prior to installation. The design readout speed for Falcon was higher, at 40 frames/s (though only 18 frames/s was so far achieved in the commercial implementation), than the 4 frames/s used for TEMAPS. Soon after the introduction

**Table 1** Specifications of the 4k x 4k TEMAPS Sensors, Prototype for Falcon I, II, and III, Designed for Use in Transmission Electron Microscopy

| Parameter | Value | Unit |
|---|---|---|
| Technology | 0.35 | $\mu$m |
| Format | 4k $\times$ 4k; i.e., 16 M pixel | |
| Pitch | 14 | $\mu$m |
| Focal plane size | 57.3 $\times$ 57.3 | mm $\times$ mm |
| Sensor size | 61 $\times$ 63 | mm $\times$ mm |
| Number of analog ports | 32 | |
| Maximum frame rate | 40 | fps |
| Radiation hardness | >500 million | 300 keV electrons |
| | About 10 | Mrad |
| Region-of-interest readout | Yes | N/A |
| Binning | x1, x2, x4 | Both directions |
| Power supply | 3.3 | V |
| Power consumption | 1,500 | mW (at full speed) |
| Gain | 6.1 | $\mu$V/e |
| Noise | 83 | e-rms (without CDS) |
| Full well (linear) | 75,000 | Electrons |
| Full well (maximum) | 120,000 | Electrons |
| Dynamic range (linear) | 59 | dB (without CDS) |
| Dynamic range (maximum) | 63.2 | dB (without CDS) |
| Quantum efficiency | 20% | At 546 nm |

*Source*: Guerrini et al. (2011b).

**Figure 5** The commercial version of the 4k × 4k detector, Falcon (http://www.fei.com/), ready for mounting in the electron microscope. (See the color plate.)

of Falcon I, a back-thinned version of the Falcon II detector, with a sensor thickness of about 50 $\mu$m and improved DQE at all spatial frequencies, was introduced. More recently, Falcon III, which is further back-thinned to about 30 $\mu$m, with even better DQE at all spatial frequencies, has been introduced. The results shown in the section "Comparison of Three Commercial Direct Detectors for Cryo-EM," later in this chapter, were obtained with Falcon II (except for Figure 12, which shows a comparison between the performance of Falcon II and Falcon III). The LDRD-RH detector has been further developed into a commercial version by Gatan called K2 Summit, which includes a counting mode—for more information, see the next section and some of the results in the "Comparison of Three Commercial Direct Detectors for Cryo-EM" section (http://www.gatan.com/contact/). Finally, the development at UCSD has been commercialized by Direct Electron, which sells a range of detectors. Results from their DE-20 detector are shown in the section "Comparison of Three Commercial Direct Detectors for Cryo-EM" (http://www.directelectron.com/).

## 3.5 Improvements in MTF and DQE Due to Electron Counting

One of the primary causes of reduction in DQE is the variability of the energy deposited per electron in the sensor. The energy deposited for each event follows a Landau distribution, which has a mean value but with a long tail; i.e., there are some electrons that deposit many times more energy than the mean value. An example of this distribution is shown in Figure 6, which is discussed in more detail next. It was suggested (Battaglia, Contarato, Denes, & Giubilato, 2009; McMullan, Clark, Turchetta, & Faruqi, 2009) that it would be possible to reduce or eliminate the effects of variability in the energy deposited by normalizing the output to 1 (i.e., by counting

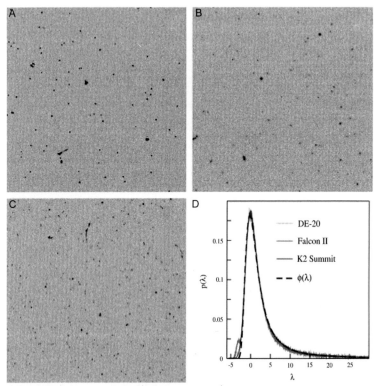

**Figure 6** Single-electron events taken from 256 x 256 pixel areas at 300 keV from three detectors: (A) DE-20, (B) Falcon II, and (C) K2 Summit. The Landau plots for all three detectors are shown in (D), along with the expected normalized Landau plot. For details, see McMullan et al. (2014). (See the color plate.)

electrons). This has the effect of assigning the same weight, unity, to recorded electrons regardless of the amount of energy deposited.

Experimentally determined Landau curves at 300 keV from such events for three commercially available detectors are shown in Figure 6 (McMullan et al., 2014). Data for the Landau distributions were derived from single-electron events recorded on the Gatan K2 Summit (http://www.gatan.com/contact/), Direct Electron DE20 (http://www.directelectron.com/), and Falcon II (http://www.fei.com/). Since it is not practical to disentangle information from more than one electron within a given area for an event, it is essential to reduce the flux of electrons to a level where the probability of more than one electron arriving in a given pixel during each frame is very low (in practice, less than 1 in about 100 pixels). As images have to be built up from a much larger number of frames, the frame rate has to be high enough to obtain an image without an excessively long exposure time (McMullan, Clark, Turchetta, & Faruqi, 2009; McMullan et al., 2011; Battaglia, Contarato, Denes, & Giubilato, 2009). Two solutions have been proposed for circumventing the problem due to variability in energy deposited by electrons, both relying on recording individual electrons, discussed next.

It was estimated by Battaglia, Contarato, Denes, and Giubilato (2009) that for 10-$\mu$m pixels, the cluster size for each single-electron event is typically 4–5 pixels, with 45% of the charge in the central (or seed) pixel; they used a centroiding method to obtain the impact position of the incident electron as accurately as possible (Battaglia, Contarato, Denes, & Giubilato, 2009) and concluded that cluster imaging improved the point spread function by a factor of 2 over an analog readout method. The DQE of this prototype was not given, but the DQE of the commercial version from Gatan, which employs counting (K2 Summit) has been measured along with two other noncounting detectors (McMullan et al., 2014). Although this method (i.e. centroiding) is widely used in X-ray detection, it can be less suitable for electron detection, as the charge deposited by an incident electron is more spread out and asymmetric, particularly in the case of thicker epilayers or smaller pixels.

A number of alternative approaches to single-electron image processing were tried by McMullan, Clark, Turchetta, and Faruqi (2009); McMullan et al. (2011) to establish the optimum mode of analysis for obtaining the impact location of the primary electron that would yield the highest DQE over a range of spatial frequencies. When using peak position or weighted centroid, DQE was improved at low spatial frequencies, but not at Nyquist frequency; i.e., the method was partially successful. The most successful approach, giving higher DQE at all frequencies, was obtained as follows: Having identified a seed pixel (pixel with maximum signal), charge was summed from adjacent pixels with a signal above a preset threshold and renormalized to unity, resulting in an effective probability distribution, which was used to calculate the impact location of the incident electron. The improvement in MTF($\omega$) and DQE($\omega$) using counting and compared to analog readout methods is shown in Figure 7. Figure 7(A) shows the improvement in MTF($\omega$) plotted as a function spatial frequency for counting mode compared to analog mode, and Figure 7(B) shows a similar improvement in DQE($\omega$) as a function of spatial frequency. The significant increase in DQE($\omega$) measured in the two types of counting modes is due to the lower-noise power contribution (NPS($\omega$)) at higher frequencies. This results in a higher DQE, especially at higher frequencies.

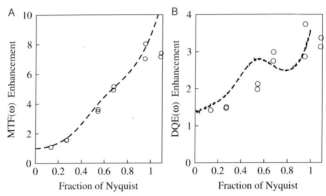

**Figure 7** Improvements in MTF and DQE as a function of spatial frequency due to renormalized single-event imaging (McMullan, Clark, Turchetta, & Faruqi, 2009).

A slightly different approach was used by Battaglia, Contarato, Denes, and Giubilato, (2009) for single event processing. In their case, they had CMOS sensors with 10-$\mu$m pixels. Using simulations with Geant 4 (http://geant4.cern.ch/), they estimated that an event would occupy fewer than 5 pixels, with 45% of the charge in the central pixel. After identifying the seed pixel with the largest signal in an event, they fitted a Gaussian distribution to the event cluster and used the centroid of the cluster. This procedure, previously used for locating the impact of X-ray photons in multiwire gas detectors (Faruqi, 1988) at a finer resolution than the pixel pitch, effectively increases the spatial resolution of the detector. This improvement was reflected in images obtained in counting mode. A similar algorithm for locating the centroid of single-electron events, K2 Summit, was implemented in the commercial version of this detector, when operating in the super-resolution mode, by Gatan (http://www.gatan.com/contact/). In order to cope with the reduction of counting rates in a frame and to keep the total exposure time sufficiently short, the internal readout is designed to be 400 frames/s. In super-resolution mode, the centroid is calculated to sub-pixel accuracy, and this is the number that is stored and used in the image. Considerable improvements in the DQE at all spatial frequencies are shown in plots comparing the DQE as a function of spatial frequency for analog (integrating-mode) readout and single-electron counting mode readout by about a factor of 2, as expected from test results (McMullan et al., 2014). Further details are given in comparative measurements of the DQE for three commercial detectors, including K2 used in counting mode, in the next two sections of this chapter.

## 3.6 Improved DQE of the K2 Detector with Electron Counting Compared to Analog Readout

The first commercially available direct electron detector with counting capabilities was the K2 Summit, marketed by Gatan. K2 Summit was based on the previous work done by Gatan and at LBNL, which was discussed in the section "Improvement in MTF and DQE Due to Electron Counting," earlier in this chapter (also see Battaglia, Contarato, Denes, & Giubilato, 2009). The detector has an internal readout speed of 400 frames/s, with external output of 40 frames/s, which are produced after electronic processing to obtain the event centroids. The improvement in DQE as a function of spatial frequency for electron counting as compared to analog readouts is shown in Figure 8(A) (Li et al., 2013). The plot shows that the DQE in counting mode is considerably higher at all spatial frequencies compared to analog readouts and extends well beyond the Nyquist frequency. For comparison, DQE data from a Gatan US4000 Ultrascan CCD camera is also shown, which highlights the improvements in DQE, as with all CMOS-based detectors, with or without electron counting options. As expected, some counting losses occur with higher counting rates, and this effect is shown in Figure 8(B) (Li et al., 2013).

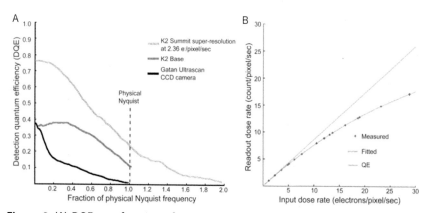

**Figure 8** (A) DQE as a function of spatial frequency for K2 Summit in linear (analog) mode, counting mode, and comparison with an Ultrascan CCD. (B) Relation between input and output rates, which have a linear relation up to 5 electrons/pixel/s, but some losses occur at higher counting rates (Li et al., 2013). Used with permission from Nature Methods. (See the color plate.)

## 3.7 Comparison of Three Commercial Direct Detectors for Cryo-EM

Comparative tests on three commercially available direct detectors for cryo-EM have been reported by McMullan et al. (2014); a summary of the results is included here, but for further details regarding the measurements., the original publication should be consulted. The tests included the most relevant properties useful in electron microscopy (namely MTF, NPS, and DQE) as a function of spatial frequency at 300 keV; the measurements of these parameters were done under similar conditions. The detectors tested were Direct Electron DE-20 (http://www.directelectron.com/), FEI Falcon II (http://www.fei.com/), and the Gatan K2 Summit (http://www.gatan.com/contact/). DE-20 contains approximately 5k x 4k 6.4-$\mu$m pixels, capable of reading out at about 32 frames/s and is based on the developments on direct electron detectors carried out originally at the UCSD (Milazzo et al., 2005, 2011). The K2 Summit contains just under 4k x 4k 5-$\mu$m pixels, capable of an internal readout at 400 frames/s and is adapted from developments carried out at the LBNL for the TEAM project (Battaglia et al., 2009; Battaglia, Contarato, Denes, & Giubilato, 2009). Finally, Falcon II is based on developments carried out at the Rutherford Appleton Laboratory, Oxford, contains 4k $\times$ 4k pixels of size 14 $\mu$m, and is capable of reading out at 18 frames/s (Guerrini et al., 2011a, b).

The main findings of the tests are summarized in Figures 9, 10, and 11, which show the variation of MTF, NPS, and DQE as a function of spatial frequency for the three detectors, along with DQE for film in the DQE plots. The plots on the noise power spectrum (NPS) in Figure 10 also contain plots of $(MTF)^2$ (from Figure 9) as a dashed line; these $MTF^2$ plots are included here because the ratio of $(MTF)^2$ to NPS is proportional to the DQE. Finally, the DQE of the three detectors and for film as a function of spatial frequency at 300 keV is shown in Figure 11, reproduced from McMullan et al. (2014). For these measurements, K2 was operated in a counting mode with sufficiently low count rates (about 1 electron/pixel/s) to avoid losses due to double counting [see Figure 8(B)]; Falcon II and DE-20 were operated in an analog (integrating) readout mode, also with the optimal dose for these detectors. The DQE for the K2 was higher than all detectors for low spatial frequencies; but at higher spatial frequencies (above half-Nyquist frequency), the DQE for Falcon II is slightly higher and Falcon III significantly higher (Figure 12). All three detectors were superior to film at all spatial frequencies. The added bonus of being able to record movie-type images with direct

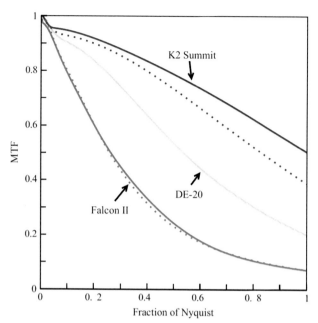

**Figure 9** Experimental plots for MTF for the three detectors. DE-20 is shown in green; K2 Summit shown in blue, with super-resolution counting mode (shown as a line) and normal counting mode (shown as dots); Falcon II is shown as red line with derived MTF from NPS shown as dots (McMullan et al., 2014). (See the color plate.)

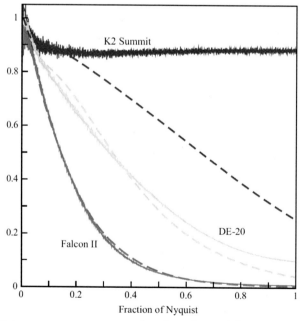

**Figure 10** Noise power spectrum as a function of spatial frequency for K2 Summit shown in blue, DE20 shown in green, and Falcon II shown in red. Dashed lines represent $MTF^2$ measurements (McMullan et al., 2014). (See the color plate.)

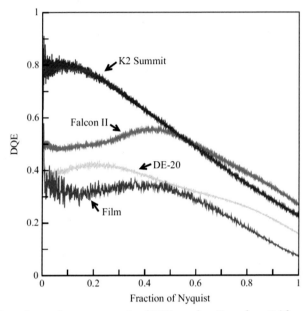

**Figure 11** Experimental measurements of DQE as a function of spatial frequency for K2 Summit in super-resolution; counting mode in blue, Falcon II in red, DE-20 in green, and film in black (McMullan et al., 2014). (See the color plate.)

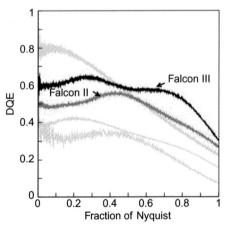

**Figure 12** DQE as a function of spatial frequency at 300 keV for the back-thinned Falcon II with a sensor thickness of about 50 $\mu$m, and the further back-thinned Falcon III with a thickness of about 30 $\mu$m, showing improved DQE at all spatial frequencies. (See the color plate.)

detectors to cope with specimen movements due to beam-specimen interaction makes these devices even more attractive.

## 4. IMAGING PROBLEMS AND SOLUTIONS PROVIDED BY DIRECT DETECTORS

Although very successful in helping to generate a large number of macromolecular structures to medium (blob) resolution, one of the most important limitations of single-particle imaging until recently has been the inability to obtain structural information to atomic resolution (Frank, 2009; Cheng & Walz, 2009; Grigorieff & Harrison, 2011). According to calculations based on the physics of the scattering process, it was estimated (Henderson, 1995; Glaeser & Hall, 2011) that it should be possible to obtain near-atomic resolution structures of large macromolecules with single-particle imaging by averaging a given number of images. For a large molecule [say, a ribosome (3.3 MD)], it should be possible to obtain near-atomic resolution by averaging about 10,000 particle images, provided that their orientation was established and they were accurately aligned prior to averaging. Experimentally, results have fallen well short of such expectations; and a number of reasons have been advanced to explain the shortfall between theory and experiment, particularly beam-induced specimen motion during exposure, which leads to image blurring and loss of contrast (Henderson & Glaeser, 1985; Glaeser, McMullan, Faruqi, & Henderson, 2011; Henderson, 1992; Henderson & McMullan, 2013). How we can bridge the gap between expectation and practical reality with the help of direct detectors is the topic of this section.

### 4.1 Early Prototype Experiments on Single Particles with a Direct Detector (Medipix2)

As described previously (Faruqi et al., 2005b), Medipix2, based on hybrid pixel technology with a noiseless readout, allowed summation or averaging of successive images without increasing noise (McMullan & Faruqi, 2008). The primary aim of the experiment was to investigate the feasibility of correcting for specimen movement during imaging, regardless of the causes of movement. In fact, the main cause of movement in this case was the stage drift in the microscope (Philips CM 12), as the total exposure for an image was 65 seconds (i.e., very long), with data collected in 65 1-s frames. The sensor in Medipix2 consists of a 300-$\mu$m-thick silicon layer, adequate for generating a large signal from 120-keV electrons and sufficiently thick to

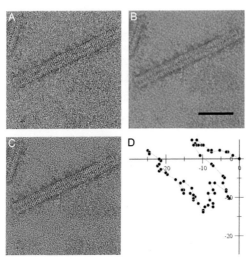

**Figure 13** Early attempts at alignment of images of negatively stained TMV stacked disks. (A) First (single) image, (B) sum of 65 images without alignment, (C) sum of 65 images with alignment, (D) shift in Å inferred from the alignment algorithm (McMullan & Faruqi, 2008).

prevent the incident electrons at this energy from reaching the radiation-sensitive electronics chip located under the sensor layer. The specimen [tobacco mosaic virus (TMV) coat protein stacked discs] was negatively stained and not susceptible to radiation damage. The TMV discs were imaged in 65 × 1 s images with a dose of 80 electrons/pixel, in what would, in contemporary terms, be called *movie-mode imaging*. Due to the fairly large stage drift [shown in Figure 13(D)], the sum of the 65 images looks quite blurred in Figure 13(B), much worse than the single image in Figure 13(A). However, when the individual images were aligned prior to averaging, the result is a much sharper image [shown in Figure 13(C)]. The amount of image shift required for aligning the images, equivalent to the inferred sample movement during the time of the exposure in Å, is shown in Figure 13(D).

## 4.2 First Experiments on Beam-Induced Movement with a Direct Detector (Medipix2)

As mentioned at the beginning of this section, it has been recognized for many years that poor contrast in cryo-images is largely caused by beam-induced specimen movement, which leads to image blurring (Henderson & Glaeser, 1985). Indirect detectors, such as fiber-optically coupled CCDs, have been used in the past to investigate the effect of reducing the incident electron dose on image contrast, in order to establish a possible threshold below which movement was eliminated, in a procedure known as *dose fractionation* (Typke, Gilpin, Downing, & Glaeser, 2007). Despite reducing the electron dose to 30 times less than that used in cryo-imaging, no significant improvements in image contrast were measured; it was suggested that a further reduction in dose was required to establish a possible lower dose limit for preventing specimen movement (Typke et al., 2007). However, due to the inadequate signal-to-noise ratio intrinsic to CCD detectors, they could not be used for these experiments; and the following experiments were done with a direct detector by Glaeser et al. (2011), who used Medipix2 (Faruqi et al., 2005b; McMullan et al., 2007). The absence of noise allowed imaging at doses many orders of magnitude below those used in normal cryo-imaging. When imaging from the same area of paraffin monolayer crystals with the dose reduced to only 500 electrons per frame, a sequence of 400 frames was recorded in a movie format. Despite the ultralow dose, the image contrast was still found to be lower than expected, presumably due to specimen movement. It is worth adding that perfect images from paraffin were obtained in the same series of experiments, but only by using a much thicker carbon support, which prevented beam-induced motion, but which unfortunately also caused an increase in background noise if used for serious cryo-EM data collection of ice-embedded specimen (Glaeser et al., 2011).

## 4.3 Following the Beam-Induced Movement of a Large Virus with a Direct Detector

One of the first attempts to study the amount of beam-induced specimen movement in detail using a direct detector was made by Brilot et al. (2012), who based their experiments on an ice-embedded large virus with a previously known structure—namely, the 70-MDa rotavirus double-layer particle (DLP) (Zhang et al., 2008). Beam-induced movement was studied with a back-thinned direct detector, the DE12, with 4k × 3k pixels, which was capable of reading out 40 frames/s; the DE-12 preceded the development of the DE-20, which was used for the comparative tests discussed in the section entitled "Comparison of Three Commercial Direct Detectors for Cryo-EM," earlier in this chapter (http://www.directelectron.com/). Keeping the incident electron dose at a fairly low level (0.5 electron/ $Å^2$/frame), the authors were able to investigate a number of parameters relevant for obtaining high-resolution images, such as the amount of translational and orientational shift in the DLPs and the time course of those changes during the exposure. Applying linear shifts to images, they were able to obtain much sharper images, with the clear implication that if this technique were applied to images of unknown single particles, improved resolution would be achievable. By measuring the time course of the movements during exposure, they also had further suggestions for improving resolution: motion was worst at the start of the exposure; for example, the first 10 e/$\mathring{A}^2$ causes far more change in orientation (2.5°) than the following 20 e/$\mathring{A}^2$ (1°). This suggests different possible strategies for imaging: by dividing the exposure into finer time intervals, it might be possible to accurately track the movement of the particles. Shorter exposures inevitably lead to weaker images and a stronger requirement for detectors with high DQE.

In a further study aimed at investigating the causes of image blurring (Campbell et al., 2012), using the same direct detector DE12 (http://www.directelectron.com/), produced movies during an exposure of the rotavirus DLP. They collected 16 frame movies and made 4 frame averages, each containing 8 electrons/$\mathring{A}^2$, to study movements during imaging due to beam-induced effects or stage drifts. They found that the movement, after an initial fast phase, slowed down. An area of the image containing DLPs with about 10 $\mathring{A}$ shifts and 1° changes in orientation during imaging is reproduced in Figure 14. If the correct lateral shifts were applied to the image, it enhanced the Thon ring pattern from that shown in Figure 14(C) to (D), demonstrating that motion-corrected maps have an improved resolution of 4.4 $\mathring{A}$ compared to 4.9 $\mathring{A}$ for maps without motion correction. The authors note that the 4 $\mathring{A}$ map required more than 500,000 subunits averaged, whereas, according to theoretical predictions (Henderson, 1995), this number should be about two orders of magnitude less; i.e., further improvements are needed, some of which are discussed in the following section.

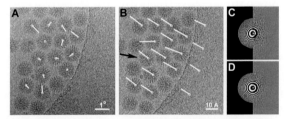

**Figure 14** Correction for image blurring due to rotation and movement in rotavirus DLPs during exposure by movie frame alignment; from Campbell et al. (2012), by permission. (A) Vectors denote the angular difference between first four frames and last four frames. (B) Vectors showing amount of lateral motion and final image after correction for movement. (C) Thon rings calculated from uncorrected images. (D) Thon rings after frame alignment showing higher resolution.

## 4.4 Processing of Images Collected in Movie Mode

As described in the "Comparison of Three Commercial Direct Detectors for Cryo-EM," section, a movie-mode readout option is available for all three direct detectors considered in this discussion (namely, K2 Summit, Falcon, and the Direct Electron detector systems). To take advantage of dose-fractionated movie mode imaging, new types of data processing processing was required, and we have included three examples based on data obtained with K2 Summit and Falcon. K2 Summit is capable of reading out 40 frames/s, and the Falcon 18 frames/s. For these two detectors, we have selected three examples of software processing, which illustrate the different approaches to the problem along with some biological results showing the improvements obtained in structure determination due to these approaches.

A powerful strategy for beam-induced motion correction (or due to other causes, such as stage instability) has been developed by Li et al. (2013) using the K2 Summit detector (http://www.gatan.com/contact/). The detector offers a high signal-to-noise ratio; i.e., excellent DQE at all spatial frequencies due to electron counting, as already discussed in the section entitled "Improved DQE of the K2 Detector with Electron Counting Compared to Analog Readout," earlier in this chapter. The detector has an external readout of 40 frames/s. Correcting for motion involved dividing the total exposure, typically from a dose of 20–30 electrons/$\text{Å}^2$, into a stack of subframes (i.e., collecting dose-fractionated images), with a sufficiently short subframe period to prevent large movements within the subframe. The total image shift was considered to be the sum of a number of linear shifts between adjacent subframes. The value of the shift between successive subframes is obtained from the peak value of the cross-correlation between the two frames. Alignment of subframes is made possible due to the extremely low readout noise in the detector, and image averaging after alignment produced dramatic improvements in image contrast compared with nonaligned images.

The motion-correction method was applied to a frozen-hydrated (test) specimen, the 20S proteasome (700 kDa molecular weight, D7 symmetry), whose structure has previously been determined to be 3.4 Å by X-ray diffraction (Lowe et al., 1995). Before motion correction was done, the resolution obtained was 4.2 Å, which was already better than the previous best resolution obtained with cryo-EM using film (Rabl et al., 2008). Motion correction improved the resolution to a best value of 3.4 Å, which was obtained by only using subframes 3–15 from a total of 24 frame exposures; the first two frames show the highest movement velocity, which would cause more blurring. The last frames, 16–24, display worse resolution, presumably due to higher radiation damage.

In our second example, Bai et al. (2013) employed a Bayesian method in their movie software, which takes advantage of the fact that there can only be small changes between subframes; i.e., particles may undergo small translations or rotations. The signal-to-noise ratio is high enough to see individual particles, but averaging a few adjacent frames provides sufficient contrast to be able to track individual particles during the exposure. The movie processing software was incorporated within the RELION software for analysis of single-particle structure (Scheres, 2015).

The specimens that were used to test the new software and the capabilities of the detector were the prokaryotic (MW 2.8 MDa) and eukaryotic ribosomes (MW 4 MDa); both specimens are stable and produce high-contrast images. The electron dose was kept at a very low level during data acquisition in movie mode. Movies with 16 frames during a total exposure of 1 s were collected, with the electron dose limited to 1 electron/$Å^2$/frame (i.e., a total dose of 16 electrons/ $Å^2$). The authors used the following procedure for further investigating optimum strategies for data collection for obtaining the best resolution. They subdivided the exposure into averages of 2x8, 3x6, 4x4, and 8x2 movie frames for each particle and calculated the accuracy of alignment by using Rosenthal and Henderson's tilt-pair method (Rosenthal & Henderson, 2003). They found that the accuracy of alignment improves when fewer frames were averaged, presumably due to less beam-induced reorientation occurring due to less radiation damage. However, averaging fewer frames also produces more noisy images, so there is an optimum in the degree of binning. In the case of averaging over four frames, they reported total rotations during the movie of $1.7° \pm 1.2°$ and translations of $4.2 \pm 2.3$ Å.

Overall, the authors achieved impressive results. For instance, the structure of the 80S ribosome was obtained at 4 Å resolution by averaging about 10,000 particles, whereas previously, the highest resolution using film, 5.5 Å, could be attained only after averaging 1.4 million particles (Armache et al., 2010)—i.e., about two orders of magnitude more than were used for obtaining higher resolution with the Falcon direct detector.

The third example of movie processing software, based on tracking particles, is taken from (Vinothkumar, McMullan, & Henderson, 2014). Using the Falcon detector (http://www.fei.com/), the authors recorded 89 images of β-galactosidase along with individual movie frames at a data rate of 18 frames/s. Exposures were typically 2 to 5 s (dose: 34–80 electrons/$Å^2$). Individual particles were tracked using a noise-weighted cross-correlation

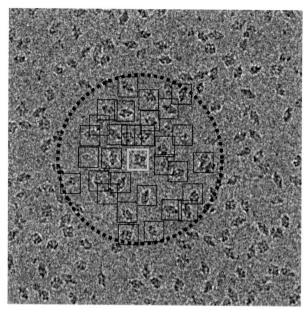

**Figure 15** Image of single particles of β-galactosidase illustrating the tracking algorithm used by Vinothkumar, McMullan, and Henderson (2014). The dotted circle indicates the region from which a group of particles (square box) was used for tracking. (See the color plate.)

function. As β-galactosidase is relatively small, it was not possible to estimate the cross-correlation peak from single frames. By using a running average of up to 11 frames, the signal-to-noise ratio was sufficient to obtain a peak consistently. Neighboring particles within a 1,000-Å radius from the center of the particle, shown as a dotted circle in Figure 15, move together with the latter and were used in the tracking software. Squares indicate the area from which the signal is used. The alignment of the particles gave an improvement in the 3D map from 6.8 to 6.3 Å. 3D maps computed from the first three frames (at 1 electron/$Å^2$) gave a worse resolution, with less information at 7 Å, compared to frames 4–9, which had the most information. After applying resolution-dependent weighting to each frame, the 3D map was improved to 5.4 Å.

## 5. HIGH-RESOLUTION STRUCTURES OBTAINED RECENTLY WITH DIRECT DETECTORS

As mentioned previously, two of the great advantages of direct detectors are that they provide much higher DQEs at all spatial frequencies

**Table 2** A Selection of the High-Resolution Structures Solved by the Use of Direct Detectors, Including Molecular Weight, Detector Used, Best Resolution Obtained, and the Source

| Specimen | Molecular Weight (MDa) | Detector Used | Resolution(Å) | Reference |
|---|---|---|---|---|
| Rotavirus | 35 | Direct Electron | 4.4 | (Campbell et al., 2012) |
| 80S ribosome | 4 | Falcon | 4.5 | (Bai et al., 2013) |
| 20S proteasome | 0.7 | K2 | 3.3 | (Li et al., 2013) |
| β-galactosidase | 0.45 | Falcon | 6 | (Chen et al., 2013) |
| TRPV1 | 0.3 | K2 | 3.4 | (Cao, Liao, Cheng, & Julius, 2013) |
| β-galatosidase + $F_v$ antibodies | 0.56 | Falcon | 5.4 | (Vinothkumar, McMullan, & Henderson, 2014) |
| TRPV + ligands | 0.3 | K2 | 3.4 | (Liao, Cao, Julius, & Cheng, 2013) |
| *Sulfolobus* turreted icosahedral virus | 75 | DE-12 | 4.5 | (Campbell et al., 2012) |
| F420-reducing hydrogenase (Frh) | 1.2 | Falcon | 3.4 | (Matteo Allegretti, McMullan, Kühlbrandt, & Vonck, 2014) |
| Mitoribosome | 1.6 | Falcon | 3.4 | (Voorhees, Fernández, Scheres, & Hegde, 2014) |
| Ribosome large subunit | 1.6 | Falcon | 3.4 | (Brown et al., 2014) |
| 80S ribosome + emitene | 4.2 | Falcon | 3.2 | (Wong et al., 2014) |
| Ryanodine receptor | 3 | K2 | 4.8 | (Zalk et al., 2015) |
| β-galactosidase | 0.56 | K2 | 3.2 | (Bartesaghi, Matthies, Banerjee, Merk, & Subramaniam, 2014) |
| Complex 1 | 1 | Falcon | 5.0 | (Vinothkumar, Zhu, & Hirst, 2014) |

**Table 2** A Selection of the High-Resolution Structures Solved by the Use of Direct Detectors, Including Molecular Weight, Detector Used, Best Resolution Obtained, and the Source—cont'd

| Specimen | Molecular Weight (MDa) | Detector Used | Resolution(Å) | Reference |
|---|---|---|---|---|
| Gamma secretase | 0.17 | Falcon | 4.5 | (Lu et al., 2014) |
| Brome mosaic virus | 3.8 | DE-12 | 4.6 | (Wang et al., 2014) |
| Dengue virus | 50 | Falcon | 4.1 | (Kostyuchenko, Chew, Ng, & Lok, 2014) |
| Ryanodine receptor | 2.2 | Falcon | 3.8 | (Yan et al., 2015) |
| TmrAB | 0.135 | K2 | 8.2 | (Kim et al., 2015) |
| Mammalian mitoribosome | 1.6 | Falcon | 3.4 | (Greber et al., 2014) |

compared to all other types of detectors, and further, due to the faster readout, have the ability to collect dose-fractionated data in movie mode. These properties have generated a large number of high-resolution structures published within the past year or so, with many still in the pipeline waiting to be printed. Due to this rapid increase in published structures, it would be difficult to give a complete list of all structures obtained with direct detectors to high resolution; instead, we have chosen a representative sample illustrating the huge amount of progress that has been made, inevitably overlapping partially with those from other recent reviews (e.g., Bai et al., 2015).

# 6. FUTURE PROSPECTS FOR FURTHER IMPROVEMENTS IN DETECTOR TECHNOLOGY AND PERFORMANCE

In this chapter, we have shown the dramatic progress made with the help of direct detectors in the elucidation of high-resolution macromolecular structures in cryo-EM. The detectors are far from perfect, and the most important detector parameter, which could be improved for all direct detectors, is the DQE at all spatial frequencies. This absolutely requires the use of counting modes, since Falcon III is already an almost perfect integrating detector. With higher (counting) DQE detectors, it should be possible to obtain higher resolutions, study lower-molecular-weight particles, or require fewer particles to achieve the same resolution. In the case of K2, which is the only currently available detector that has a fast enough frame rate to be usable for practical data collection in counting mode, the DQE at both low and high spatial frequencies could be improved. Improvements in DQE can be achieved by improving the overall signal-to-noise ratio for single-electron events. This will improve the DQE(0) since fewer events will be missed because they are below the noise level. It will also improve DQE at higher frequencies since the localization of the events will be determined more accurately in the presence of lower noise levels. In addition, it may be possible to improve the protocols used in counting methods, resulting in improved localization of the point of impact of the primary electron. Faster readout would also be useful in creating a tool to help in understanding the details of beam-induced specimen movement.

## ACKNOWLEDGMENTS

We are grateful to Tony Crowther and K. R. Vinothkumar for helpful suggestions and comments and Nigel Unwin for a detailed critique of the manuscript.

## REFERENCES

Adrian, M., Dubochet, J., Lepault, J., & McDowall, A. W. (1984). Cryo-electron microscopy of viruses. *Nature, 308,* 32–36.

Agard, D., Cheng, Y., Glaeser, R. M., & Subramaniam, S. (2014). Single-particle cryo-electron microscopy (cryo-EM): Progress, challenges, and perspectives for further improvement. *Advances in Imaging and Electron Physics, 185,* 113–137.

Anaxagoras, T., Kent, P., Allinson, N., Turchetta, R., Pickering, T., Maneuski, D., et al. (2010). eLeNA: A parametric CMOS active-pixel sensor for the evaluation of reset noise reduction architectures. *IEEE Transactions on Electron Devices, 57,* 2163–2175.

Armache, J.-P., Jarasch, A., Anger, A. M., Villa, E., Becker, T., Bhushan, S., et al. (2010). Cryo-EM structure and rRNA model of a translating eukaryotic 80S ribosome at 5.5-Å resolution. *Proceedings of the National Academy of Sciences, 107,* 19748–19753.

Arndt, U. W. (1986). X-ray position-sensitive detectors. *Journal of Applied Crystallography, 19,* 145–163.

Bai, X.-C., Fernandez, I., McMullan, G., & Scheres, S. (2013). Ribosome structures to near-atomic resolution from thirty thousand cryo-EM particles. *eLife, 2,* e00461.

Bai, X.-C., McMullan, G., & Scheres, S. H. W. (2015). How cryo-EM is revolutionizing structural biology. *Trends in Biochemical Sciences, 40,* 49–57.

Bammes, B. E., Rochat, R. H., Jakana, J., & Chiu, W. (2011). Practical performance evaluation of a 10k x 10k CCD for electron cryo-microscopy. *Journal of Structural Biology, 175,* 384–393.

Bartesaghi, A., Matthies, D., Banerjee, S., Merk, A., & Subramaniam, S. (2014). Structure of β-galactosidase at 3.2-Å resolution obtained by cryo-electron microscopy. *Proceedings of the National Academy of Sciences, 111,* 11709–11714.

Battaglia, M., Contarato, D., Denes, P., Doering, D., Duden, T., Krieger, B., et al. (2010). Characterisation of a CMOS active pixel sensor for use in the TEAM microscope.

*Nuclear Instruments and Methods in Physics Research Section A: Accelerators, Spectrometers, Detectors and Associated Equipment, 622,* 669–677.

Battaglia, M., Contarato, D., Denes, P., Doering, D., Giubilato, P., Kim, T. S., et al. (2009). A rad-hard CMOS active pixel sensor for electron microscopy. *Nuclear Instruments and Methods in Physics Research A, 598,* 642–649.

Battaglia, M., Contarato, D., Denes, P., & Giubilato, P. (2009). Cluster imaging with a direct detection CMOS pixel sensor in transmission electron microscopy. *Nuclear Instruments and Methods in Physics Research Section A: Accelerators, Spectrometers, Detectors and Associated Equipment, 608,* 363–365.

Bogaerts, J., Dierckx, B., Meynants, G., & Uwaerts, D. (2003). Total dose and displacement damage effects in a radiation-hardened CMOS APS. *IEEE Transactions on Electron Devices, 50,* 84–90.

Brilot, A. F., Chen, J. Z., Cheng, A., Pan, J., Harrison, S. C., Potter, C. S., et al. (2012). Beam-induced motion of vitrified specimen on holey carbon film. *Journal of Structural Biology, 177,* 630–637.

Brown, A., Amunts, A., Bai, X.-C., Sugimoto, Y., Edwards, P. C., Murshudov, G., et al. (2014). Structure of the large ribosomal subunit from human mitochondria. *Science, 346,* 718–722.

Campbell, M. (2011). 10 years of the Medipix2 collaboration. *Nuclear Instruments and Methods in Physics Research Section A: Accelerators, Spectrometers, Detectors and Associated Equipment, 633*(Suppl. 1), S1–S10.

Campbell, M., Anelli, G., Cantatore, E., Faccio, F., Heijne, E. H. M., Jarron, P., et al. (2001). An introduction to deep submicron CMOS for vertex applications. *Nuclear Instruments and Methods in Physics Research Section A: Accelerators, Spectrometers, Detectors and Associated Equipment, 473,* 140–145.

Campbell, M. G., Cheng, A., Brilot, A. F., Moeller, A., Lyumkis, D., Veesler, D., et al. (2012). Movies of ice-embedded particles enhance resolution in electron cryo-microscopy. *Structure, 20,* 1823–1828.

Cao, E., Liao, M., Cheng, Y., & Julius, D. (2013). TRPV1 structures in distinct conformations reveal activation mechanisms. *Nature, 504,* 113–118.

Chen, S., McMullan, G., Faruqi, A. R., Murshudov, G. N., Short, J. M., Scheres, S. H. W., et al. (2013). High-resolution noise substitution to measure overfitting and validate resolution in 3D structure determination by single-particle electron cryomicroscopy. *Ultramicroscopy, 135,* 24–35.

Cheng, Y., & Walz, T. (2009). The advent of near-atomic resolution in single-particle electron microscopy. *Annual Review of Biochemistry, 78,* 723–742.

Clare, D. K., & Orlova, E. V. (2010). 4.6 Å cryo-EM reconstruction of tobacco mosaic virus from images recorded at 300 keV on a 4k × 4k CCD camera. *Journal of Structural Biology, 171,* 303–308.

Dainty, J. C., & Shaw, R. (1974). *Image science.* Academic Press.

Deptuch, G., Besson, A., Rehak, P., Szelezniak, M., Wall, J., Winter, M., et al. (2007). Direct electron imaging in electron microscopy with monolithic active pixel sensors. *Ultramicroscopy, 107,* 674–684.

Dubochet, J., Adrian, M., Chang, J.-J., Homo, J.-C., Lepault, J., McDowall, A. W., et al. (1988). Cryo-electron microscopy of vitrified specimens. *Quarterly Reviews of Biophysics, 21,* 129–228.

Faruqi, A. R. (1988). Applications of multiwire proportional chambers to time resolved X-ray studies on muscle. *Nuclear Instruments and Methods in Physics Research Section A: Accelerators, Spectrometers, Detectors and Associated Equipment, 269,* 362–368.

Faruqi, A. R. (1991). Applications in biology and condensed matter physics. *Nuclear Instruments and Methods in Physics Research Section A: Accelerators, Spectrometers, Detectors and Associated Equipment, 310,* 14–23.

Faruqi, A. R. (2001). Prospects for hybrid pixel detectors in electron microscopy. *Nuclear Instruments and Methods in Physics Research, A466*, 146–154.

Faruqi, A. R. (2007). Direct electron detectors for electron microscopy. *Advances in Imaging and Electron Physics, 145*, 55–94.

Faruqi, A. R., & Andrews, H. N. (1997). Cooled CCD camera with tapered fibre optics for electron microscopy. *Nuclear Instruments and Methods in Physics Research Section A: Accelerators, Spectrometers, Detectors and Associated Equipment, 392*, 233–236.

Faruqi, A. R., Cattermole, D. M., Henderson, R., Mikulec, B., & Raeburn, C. (2003). Evaluation of a hybrid pixel detector for electron microscopy. *Ultramicroscopy, 94*, 263–276.

Faruqi, A. R., Henderson, R., & Holmes, J. (2006). Radiation damage studies on STAR250 CMOS sensor at 300 keV for electron microscopy. *Nuclear Instruments and Methods in Physics Research Section A: Accelerators, Spectrometers, Detectors and Associated Equipment, 565*, 139–143.

Faruqi, A. R., Henderson, R., & McMullan, G. (2013). Recent developments in direct electron detectors for electron cryo-microscopy. *Proceedings of Science (Vertex 2013)*, [Online], 044.

Faruqi, A. R., Henderson, R., Prydderch, M., Turchetta, R., Allport, P., & Evans, A. (2005a). Direct single-electron detection with a CMOS detector for electron microscopy. *Nuclear Instruments and Methods in Physics Research A, 546*, 170–175.

Faruqi, A. R., Henderson, R., & Tlustos, L. (2005b). Noiseless direct detection of electrons in Medipix2 for electron microscopy. *Nuclear Instruments and Methods in Physics Research Section A: Accelerators, Spectrometers, Detectors and Associated Equipment, 546*, 160–163.

Faruqi, A. R., & McMullan, G. (2011). Electronic detectors for electron microscopy. *Quarterly Reviews of Biophysics, 44*, 357–390.

Faruqi, A. R., & Subramaniam, S. (2000). CCD detectors in high-resolution biological electron microscopy. *Quarterly Reviews of Biophysics, 33*, 1–28.

Frank, J. (2009). Single-particle reconstruction of biological macromolecules in electron microscopy - 30 years. *Quarterly Reviews of Biophysics, 42*, 139–158.

Glaeser, R. M., & Hall, R. J. (2011). Reaching the information limit in cryo-EM of biological macromolecules: Experimental aspects. *Biophysical Journal, 100*, 2331–2337.

Glaeser, R. M., McMullan, G., Faruqi, A. R., & Henderson, R. (2011). Images of paraffin monolayer crystals with perfect contrast: Minimization of beam-induced specimen motion. *Ultramicroscopy, 111*, 90–100.

Greber, B. J., Boehringer, D., Leibundgut, M., Bieri, P., Leitner, A., & Schmitz, N. (2014). The complete structure of the large subunit of the mammalian mitochondrial ribosome. *Nature, 515*, 283–286.

Grigorieff, N., & Harrison, S. C. (2011). Near-atomic resolution reconstructions of icosahedral viruses from electron cryo-microscopy. *Current Opinion in Structural Biology, 21*, 265–273.

Guerrini, N., Turchetta, R., Van Hoften, G., Faruqi, A. R., McMullan, G., & Henderson, R. (2011a). A 61 mm x 63 mm, 16 million pixels, 40 frames per second, radiation-hard CMOS image sensor for transmission electron microscopy. In *Proceedings of the 2011 International Image Sensor Workshop, Hakodate-Onuma, Japan* (pp. 293–296).

Guerrini, N., Turchetta, R., Van Hoften, G., Henderson, R., McMullan, G., & Faruqi, A. R. (2011b). A high frame rate, 16 million pixels, radiation hard CMOS sensor. *Journal of Instrumentation, 6*, C03003.

Henderson, R. (1992). Image contrast in high-resolution electron microscopy of biological molecules. *Ultramicroscopy, 46*, 1–18.

Henderson, R. (1995). The progress and limitations of neutrons, electrons and X-rays for atomic resolution microscopy of unstained biological molecules. *Quarterly Reviews of Biophysics, 28*, 171–193.

Henderson, R. (2015). Overview and future of single-particle electron cryomicroscopy. *Archives of Biochemistry and Biophysics*, in press. http://dx.doi.org/10.1016/j.abb.2015.02.036.

Henderson, R., & Glaeser, R. M. (1985). Quantitative analysis of image contrast in electron micrographs of beam-sensitive crystals. *Ultramicroscopy, 16*, 139–150.

Henderson, R., & McMullan, G. (2013). Problems in obtaining perfect images by single-particle electron cryomicroscopy of biological structures in amorphous ice. *Microscopy, 62*, 43–50.

CERN (2015). http://geant4.cern.ch/.

Direct Electron (2015). http://www.directelectron.com/.

FEI (2015). http://www.fei.com/.

Gatan (2015). http://www.gatan.com/contact/.

TVIPS (2015). http://www.tvips.com/.

Jin, L., Milazzo, A.-C., Kleinfelder, S., Li, S., Leblanc, P., Duttweiler, F., et al. (2008). Applications of direct detection device in transmission electron microscopy. *Journal of Structural Biology, 161*, 352–358.

Kim, J., Wu, S., Tomasiak, T. M., Mergel, C., Winter, M. B., Stiller, S. B., et al. (2015). Subnanometre-resolution electron cryomicroscopy structure of a heterodimeric ABC exporter. *Nature, 517*, 396–400.

Kostyuchenko, V. A., Chew, P. L., Ng, T.-S., & Lok, S.-M. (2014). Near-atomic resolution cryo-electron microscopic structure of dengue serotype 4 virus. *Journal of Virology, 88*, 477–482.

Kühlbrandt, W. (2014). Cryo-EM enters a new era. *eLife, 3*, e03678. [Online] Available: http://elifesciences.org/elife/3/e03678.full.pdf, Accessed 13.08.14.

Li, X., Mooney, P., Zheng, S., Booth, C. R., Braunfeld, M. B., Gubbens, S., et al. (2013). Electron counting and beam-induced motion correction enable near-atomic-resolution single-particle cryo-EM. *Nature Methods, 10*, 584–590.

Liao, M., Cao, E., Julius, D., & Cheng, Y. (2013). Structure of the TRPV1 ion channel determined by electron cryo-microscopy. *Nature, 504*, 107–112.

Liao, M., Cao, E., Julius, D., & Cheng, Y. (2014). Single-particle electron cryo-microscopy of a mammalian ion channel. *Current Opinion in Structural Biology, 27*, 1–7.

Liu, W., Wacker, D., Gati, C., Han, G. W., James, D., Wang, D., et al. (2013). Serial femtosecond crystallography of G protein–coupled receptors. *Science, 342*, 1521–1524.

Llopart, X., & Campbell, M. (2003). First test measurements of a 64 K-pixel readout chip working in single-photon counting mode. *Nuclear Instruments and Methods of Physics Research, Section A, 509*, 157–163.

Llopart, X., Campbell, M., Dinapoli, R., San Secundo, D., & Pernigotti, E. (2002). Medipix2: A 64 k-pixel readout chip with 55-$\mu$m-square elements working in single-photon counting mode. *IEEE Transactions in Nuclear Science, 49*, 2279–2283.

Lowe, J., Stock, D., Jap, B., Zwickl, P., Baumeister, W., & Huber, R. (1995). Crystal structure of the 20S proteasome from the Archaeon T. acidophilum at 3.4 A resolution. *Science, 268*, 533–539.

Lu, P., Bai, X.-C., Ma, D., Xie, T., Yan, C., Sun, L., et al. (2014). Three-dimensional structure of human [ggr]-secretase. *Nature, 512*, 166–170.

Lucic, V., Forster, F., & Baumeister, W. (2005). Structural studies by electron tomography: From cells to molecules. *Annual Review of Biochemistry, 74*, 833–865.

Matteo Allegretti, D. J. M., McMullan, G., Kühlbrandt, W., & Vonck, J. (2014). Atomic model of the F420-reducing [NiFe] hydrogenase by electron cryo-microscopy using a direct electron detector. *eLife, 2014*(3), e01963.

McMullan, G., Cattermole, D. M., Chen, S., Henderson, R., Llopart, X., Summerfield, C., et al. (2007). Electron imaging with Medipix2 hybrid pixel detector. *Ultramicroscopy, 107*, 401–413.

McMullan, G., Chen, S., Henderson, R., & Faruqi, A. R. (2009). The detective quantum efficiency of electron area detectors in electron microscopy. *Ultramicroscopy, 109*, 1126–1143.

McMullan, G., Clark, A. T., Turchetta, R., & Faruqi, A. R. (2009). Enhanced imaging in low-dose electron microscopy using electron counting. *Ultramicroscopy, 109*, 1411–1416.

McMullan, G., & Faruqi, A. R. (2008). Electron microscope imaging of single-particles using the Medipix2 detector. *Nuclear Instruments and Methods in Physics Research Section A: Accelerators, Spectrometers, Detectors and Associated Equipment, 591*, 129–133.

McMullan, G., Faruqi, A. R., Clare, D., & Henderson, R. (2014). Comparison of optimal performance at 300 keV of three direct electron detectors for use in low-dose electron microscopy. *Ultramicroscopy, 147*, 156–163.

McMullan, G., Faruqi, A. R., Henderson, R., Guerrini, N., Turchetta, R., Jacobs, A., et al. (2009). Experimental observation of the improvement in MTF from backthinning a CMOS direct electron detector. *Ultramicroscopy, 109*, 1144–1147.

McMullan, G., Turchetta, R., & Faruqi, A. R. (2011). Single-event imaging for electron microscopy using MAPS detectors. *Journal of Instrumentation, 6*, C04001.

Meyer, R. R., & Kirkland, A. (1998). The effects of electron and photon scattering on signal and noise transfer properties of scintillators in CCD cameras used for electron detection. *Ultramicroscopy, 75*, 23–33.

Meyer, R. R., & Kirkland, A. I. (2000). Characterisation of the signal and noise transfer of CCD cameras for electron detection. *Microscopy Research and Technique, 49*, 269–280.

Milazzo, A.-C., Cheng, A., Moeller, A., Lyumkis, D., Jacovetty, E., Polukas, J., et al. (2011). Initial evaluation of a direct detection device detector for single-particle cryo-electron microscopy. *Journal of Structural Biology, 176*, 404–408.

Milazzo, A., Leblanc, P., Duttweiler, F., Jin, L., Bouwer, J. C., Peltier, S., et al. (2005). Active pixel sensor array as a detector for electron microscopy. *Ultramicroscopy, 104*, 152–159.

Nannenga, B. L., Shi, D., Leslie, A. G. W., & Gonen, T. (2014). High-resolution structure determination by continuous-rotation data collection in MicroED. *Nature Methods, 11*, 927–930.

Nederlof, I., van Genderen, E., Li, Y.-W., & Abrahams, J. P. (2013). A Medipix quantum area detector allows rotation electron diffraction data collection from submicrometre three-dimensional protein crystals. *Acta Crystallographica, Section D, 69*, 1223–1230.

Prydderch, M. L., Waltham, N. J., Turchetta, R., French, M. J., Holt, R., Marshall, A., et al. (2003). A 512 × 512 CMOS monolithic active pixel sensor with integrated ADCs for space science. *Nuclear Instruments and Methods in Physics Research, A512*, 358–367.

Rabl, J., Smith, D. M., Yu, Y., Chang, S.-C., Goldberg, A. L., & Cheng, Y. (2008). Mechanism of gate opening in the 20S proteasome by the proteasomal ATPases. *Molecular Cell, 30*, 360–368.

Rosenthal, P. B., & Henderson, R. (2003). Optimal determination of particle orientation, absolute hand, and contrast loss in single-particle electron cryomicroscopy. *Journal of Molecular Biology, 333*, 721–745.

Ruskin, R. S., Yu, Z., & Grigorieff, N. (2013). Quantitative characterization of electron detectors for transmission electron microscopy. *Journal of Structural Biology, 184*, 385–393.

Sander, B., Golas, M. M., & Stark, H. (2005). Advantages of CCD detectors for de novo three-dimensional structure determination in single-particle electron microscopy. *Journal of Structural Biology, 151*, 92–105.

Scheres, S. H. W. (2012). RELION: Implementation of a Bayesian approach to cryo-EM structure determination. *Journal of Structural Biology, 180*, 519–530.

Scheres, S. H. W. (2015). Semi-automated selection of cryo-EM particles in RELION-1.3. *Journal of Structural Biology, 189*, 114–122.

Szwedziak, P., Wang, Q., Bharat, T. A. M., Tsim, M., & Löwe, J. (2014). Architecture of the ring formed by the tubulin homologue FtsZ in bacterial cell division. *eLife*. http://dx.doi.org/10.7554/eLife.04601.

Turchetta, R. (2007). CMOS monolithic active pixel sensors (MAPS) for scientific applications: Some notes about radiation hardness. *Nuclear Instruments and Methods in Physics Research Section A: Accelerators, Spectrometers, Detectors and Associated Equipment, 583*, 131–133.

Turchetta, R., Berst, J. D., Casadei, B., Claus, G., Colledani, C., Dulinski, W., et al. (2001). A monolithic active pixel sensor for charged particle tracking and imaging using standard VLSI CMOS technology. *Nuclear Instruments and Methods in Physics Research, Section A, 458*, 677–689.

Typke, D., Gilpin, C. J., Downing, K. H., & Glaeser, R. M. (2007). Stroboscopic image capture: Reducing the dose per frame by a factor of 30 does not prevent beam-induced specimen movement in paraffin. *Ultramicroscopy, 107*, 106–115.

Vinothkumar, K. R., McMullan, G., & Henderson, R. (2014). Molecular mechanism of antibody-mediated activation of β-galactosidase. *Structure, 22*, 621–627.

Vinothkumar, K. R., Zhu, J., & Hirst, J. (2014). Architecture of mammalian respiratory complex I. *Nature, 515*, 80–84.

Voorhees, R. M., Fernández, I. S., Scheres, S. H. W., & Hegde, R. S. (2014). Structure of the mammalian ribosome-Sec61 complex to 3.4 Å resolution. *Cell, 157*, 1632–1643.

Wang, Z., Hryc, C. F., Bammes, B., Afonine, P. V., Jakana, J., Chen, D.-H., et al. (2014). An atomic model of brome mosaic virus using direct electron detection and real-space optimization. *Nature Communications, 5*.

Wong, W., Bai, X.-C., Brown, A., Fernandez, I. S., Hanssen, E., Condron, M., et al. (2014). Cryo-EM structure of the Plasmodium falciparum 80S ribosome bound to the anti-protozoan drug emetine. http://dx.doi.org/10.7554/eLife.03080.

Xuong, N.-H., Milazzo, A.-C., LeBlanc, P., Duttweiler, F., Bouwer, J., Peltier, S., et al. (2004). First use of a high-sensitivity active pixel sensor array as a detector for electron microscopy. In *Proc. SPIE 5301, Sensors and camera systems for scientific, industrial, and digital photography applications V* (pp. 242–249). http://dx.doi.org/10.1117/12.526021.

Yan, Z., Bai, X.-C., Yan, C., Wu, J., Li, Z., Xie, T., et al. (2015). Structure of the rabbit ryanodine receptor RyR1 at near-atomic resolution. *Nature, 517*, 50–55.

Zalk, R., Clarke, O. B., Georges, A. D., Grassucci, R. A., Reiken, S., Mancia, F., et al. (2015). Structure of a mammalian ryanodine receptor. *Nature, 517*(7532), 44–49.

Zhang, X., Settembre, E., Xu, C., Dormitzer, P. R., Bellamy, R., Harrison, S. C., et al. (2008). Near-atomic resolution using electron cryomicroscopy and single-particle reconstruction. *Proceedings of the National Academy of Sciences, 105*, 1867–1872.

CHAPTER THREE

# Electron Optics and Electron Microscopy Conference Proceedings and Abstracts: A Supplement

**Peter W. Hawkes***

CEMES–CNRS, Toulouse
*Corresponding author: e-mail address: hawkes@cemes.fr

## Contents

*Advances in Imaging and Electron Physics*, Volume 190
ISSN 1076-5670
http://dx.doi.org/10.1016/bs.aiep.2015.03.003

# 1. INTRODUCTION

In an earlier volume of these advances, a near-complete list of international, regional, and national conferences on electron microscopy was published, together with details of many related meetings (Hawkes, 2003). Here, some of those lists are brought up to date: the International Congresses on Electron Microscopy (ICEM), now the International Microscopy Congresses (IMC); the European Congresses on Electron Microscopy (EUREM), now the European Microscopy Congresses (EMC); the Asia-Pacific Congresses on Electron Microscopy (APEM), now the Asia-Pacific Microscopy Conferences (APMC); the Multinational Conferences on Electron Microscopy (MCEM), now the Multinational Congresses on Microscopy (MCM); and the Dreiländertagungen. The original paper contained lists of national society meetings. I have not thought it useful to update all these here, for much of the information can now be retrieved via Internet. Some guidance is, however, included for such major meetings as the EMAG, the Microscopy Society of America (listed in full) and the Microscopy Society of Southern Africa conferences (this choice is admittedly arbitrary); two series of Russian conferences are also included. In addition, the details of a selection of other conferences are provided, notably the ICXOM and Scanning charged-particle optics meetings.

For extensive background information, see Hawkes (2003); here, only a brief commentary and details of publication of the proceedings or extended abstracts of these meetings are provided. Links to current conferences are to be found on the websites of the International Federation of Societies for Microscopy (IFSM; http://ifsm.info), the European Microscopy Society (EMS; see www.eurmicsoc.org) and the Microscopy Society of America (MSA; www.microscopy.org); another list maintained by Petr Schauer is a very useful source since it contains information about past congresses (from 1996), as well as current meetings (www.petr.isibrno.cz/microscopy/meetings.php).

# 2. INTERNATIONAL CONGRESSES (ICEM, IMC)

In 1955, nine national societies of electron microscopy agreed to form an International Federation of Electron Microscope Societies (IFEMS): the founding members were Belgium, France, Germany, Great Britain, Japan, the Netherlands, Scandinavia (Denmark, Sweden and Norway), Switzerland, and the United States. In 1958, these were joined by the Czechoslovak, Hungarian, Italian, and Spanish societies, and the federation became known as the International Federation of Societies of Electron Microscopy (IFSEM); in 2002, the word *electron* was dropped, and the name became International Federation of Societies for Microscopy (IFSM). Since its foundation, the international and most regional meetings have been held under the aegis of this group, and the number of member societies has grown steadily. For current information, see the IFSM website (http://ifsm.info).

Delft, 1949. *Proceedings of the Conference on Electron Microscopy, Delft,* July 4–8, 1949 (Houwink, A. L., Le Poole, J. B., & Le Rütte, W. A., Eds.). Delft: Hoogland, 1950.

Paris, 1950. *Comptes Rendus du Premier Congrès International de Microscopie Electronique, Paris,* September 14–22, 1950. Paris: Editions de la. Revue d'Optique Théorique et Instrumentale, 1953. 2 vols.

London, 1954. *Proceedings of the Third International Conference on Electron Microscopy, London School of Hygiene and Tropical Medicine,* July 15–21, 1954 (Ross, R., Ed.). London: Royal Microscopical Society, 1956.

Berlin, 1958. *Vierter Internationaler Kongress für Elektronenmikroskopie, Berlin,* September 10–17, 1958, Verhandlungen (Bargmann, W., Möllenstedt, G., Niehrs, H., Peters, D., Ruska, E., & Wolpers, C., Eds.). Berlin: Springer, 1960. 2 vols. Available on SpringerLink (link.springer.com).

Philadelphia, 1962. *Electron Microscopy. Fifth International Congress for Electron Microscopy, Philadelphia,* August 29–September 5, 1962 (Breese, S. S., Ed.). New York: Academic Press, 1962. 2 vols.

Kyoto, 1966. *Electron Microscopy 1966. Sixth International Congress for Electron Microscopy, Kyoto,* August 28–September 4, 1966 (Uyeda, R., Ed.). Maruzen, Tokyo, 1966. 2 vols.

Grenoble, 1970. *Microscopie Electronique 1970. Résumés des Communications Présentées au Septième Congrès International, Grenoble,* August 30–September 5, 1970 (Favard, P., Ed.). Paris: Société Française de Microscopie Electronique, 1970. 3 vols.

Canberra, 1974. *Electron Microscopy 1974. Abstracts of Papers Presented to the Eighth International Congress on Electron Microscopy, Canberra,* August 25–31, 1974 (Sanders, J. V., & Goodchild, D. J., Eds.). Canberra: Australian Academy of Sciences, 1974. 2 vols.

Toronto, 1978. *Electron Microscopy 1978. Papers Presented at the Ninth International Congress on Electron Microscopy, Toronto,* August 1–9, 1978 (Sturgess, J. M., Ed.). Toronto: Microscopical Society of Canada, 1978. 3 vols.

Hamburg, 1982. *Electron Microscopy, 1982. Papers Presented at the Tenth International Congress on Electron Microscopy, Hamburg,* August 17–24, 1982. Frankfurt: Deutsche Gesellschaft für Elektronenmikroskopie, 1982. 3 vols.

Kyoto, 1986. *Electron Microscopy 1986. Proceedings of the XIth International Congress on Electron Microscopy, Kyoto,* August 31–September 7, 1986 (Imura, T., Maruse, S., & Suzuki, T., Eds..). Tokyo: Japanese Society of Electron Microscopy. 4 vols; published as a supplement to *Journal of Electron Microscopy, 35* (1986).

Seattle, 1990. *Electron Microscopy 1990. Proceedings of the XIIth International Congress for Electron Microscopy, Seattle,* August 12–18, 1990 (Peachey, L. D., & Williams, D. B., Eds..). San Francisco: San Francisco Press. 4 vols. See also *Ultramicroscopy 36*(1–3) 1–274 (1991).

Paris, 1994. *Electron Microscopy 1994. Proceedings of the 13th International Congress on Electron Microscopy Paris,* July 17–22, 1994 [Jouffrey, B., Colliex, C., Chevalier, J. P., Glas, F., Hawkes, P. W., Hernandez–Verdun, D., Schrevel, J., & Thomas, D., Eds. (Vol. 1); Jouffrey, B., Colliex, C., Chevalier, J. P., Glas, F., & Hawkes, P. W., Eds. (Vols. 2A and 2B); Jouffrey, B., Colliex, C., Hernandez-Verdun, D., Schrevel, J., & Thomas, D., Eds. (Vols. 3A and 3B)]. Les Ulis: Editions de Physique, 1994.

Cancún, 1998. *Electron Microscopy 1998. Proceedings of the 14th International Congress on Electron Microscopy, Cancún,* August 31–September 4, 1998 [Memorias del 14 to Congreso Internacional de Microscopía Electrónica celebrado en Cancún (México) del 31 de Agosto al 4 de Septiembre de 1998] (Calderón Benavides, H. A., & Yacamán, M.J., Eds.). Bristol and Philadelphia: Institute of Physics Publishing, 1998. 4 vols. See also *Micron. 31*(5) (2000).

Durban, 2002. *Electron Microscopy 2002. Proceedings of the 15th International Congress on Electron Microscopy, International Convention Centre, Durban,* September 1–6, 2002 [Cross, R., Richards, P., Witcomb, M., & Engelbrecht, J, Eds. (Vol. 1, Physical, Materials, and Earth Sciences), Cross, R., Richards, P., Witcomb, M., & Sewell, T., Eds. (Vol. 2, Life Sciences), and Cross, R., Richards, P., Witcomb, M., Engelbrecht, J., & Sewell, T., Eds. (Vol. 3, Interdisciplinary)]. Onderstepoort: Microscopy Society of Southern Africa, 2002.

Sapporo, 2006. *Proceedings of 16th International Microscopy Conference, "Microscopy for the 21st Century," Sapporo,* September 3–8, 2006 (Ichinose, H., & Sasaki, T., Eds.). Vol. 1, Biological and Medical Science; Vol. 2, Instrumentation; Vol. 3, Materials Science. Sapporo: Publication Committee of IMC16, 2006.

Rio de Janeiro, 2010. *Proceedings of IMC17, The 17th IFSM International Microscopy Congress, Rio de Janeiro,* September 19–24, 2010 (Solórzano, G., & de Souza, W., Eds.). Rio de Janeiro: Sociedade Brasileira de Microscopia e Microanálise, 2010.

Prague, 2014. *IMC-18,* Prague Convention Centre, September 7–12, 2014. Proceedings open access at www.microscopy.cz/proceedings/all. html, P. Hozak (Ed.).

Sydney, 2018.

## 3. REGIONAL CONGRESSES

### 3.1 European Regional Conferences on Electron Microscopy (EUREM, EMC)

Today, the European regional meetings are held under the aegis of the European Microscopy Society (EMS), which was created in 1998. Until then, they were under the responsibility of the Committee of European Societies of Electron Microscopy (CESEM), which was founded in 1976 at the sixth European conference, in Jerusalem and then renamed the Committee of European Societies of Microscopy (CESM) in 1994. For current membership details, see www.eurmicsoc.org or the IFSM website (http://ifsm.info).

Stockholm, 1956. *Electron Microscopy. Proceedings of the Stockholm Conference,* September 17–20, 1956 (Sjöstrand, F. J., & Rhodin, J., Eds.). Stockholm: Almqvist and Wiksells, 1957.

Delft, 1960. *Proceedings of the European Regional Conference on Electron Microscopy, Delft,* August 29–September 3, 1960 (Houwink, A. L., & Spit, B. J., Eds..). Delft: Nederlandse Vereniging voor Elektronenmicroscopie, n.d. 2 vols.

Prague, 1964. *Electron Microscopy 1964. Proceedings of the Third European Regional Conference, Prague,* August 26–September 3, 1964 (Titlbach, M., Ed.). Prague: Publishing House of the Czechoslovak Academy of Sciences, 2 vols.

Rome, 1968. *Electron Microscopy 1968. Pre-Congress Abstracts of Papers Presented at the Fourth Regional Conference, Rome,* September 1–7, 1968 (Bocciarelli, D.S., Ed.). Rome: Tipographia Poliglotta Vaticana, 2 vols.

Manchester, 1972. *Electron Microscopy 1972. Proceedings of the Fifth European Congress on Electron Microscopy, Manchester,* September 5–12, 1972. London: Institute of Physics 1972. Full versions of the contributions to the symposia on image processing and on computer–aided design in electron optics are available in *Image Processing and Computer-Aided Design in*

*Electron Optics* (Hawkes, P. W., Ed.; Academic Press, London and New York 1973). The symposium on high-voltage electron microscopy (sponsored by the Royal Microscopical Society) is recorded in full in The High-Voltage Symposium at EMCON 1972, *Journal of Microscopy,* 97(1/2), 1–268 (1973), P. R. Swann (Ed.).

Jerusalem, 1976. *Electron Microscopy 1976. Proceedings of the Sixth European Congress on Electron Microscopy, Jerusalem,* September 14–20, 1976 [Brandon, D. G., Ed. (Vol. I) and Ben-Shaul, Y., Ed. (Vol. II)]. Jerusalem: Tal International, 2 vols.

The Hague, 1980. *Electron Microscopy 1980. Proceedings of the Seventh European Congress on Electron Microscopy, The Hague,* August 24–29, 1980 [Brederoo, P., & Boom, G., Eds. (Vol. I), Brederoo, P., & Priester, W. de, Eds. (Vol. II), Brederoo, P. & Cosslett, V. E., Eds. (Vol. III), and Brederoo, P. & Landuyt, J. van, Eds. (Vol. IV)]. Vols. I and II contain the proceedings of the Seventh European Congress on Electron Microscopy, Vol. III those of the Ninth International Conference on X-Ray Optics and Microanalysis, and Vol. IV those of the Sixth International Conference on High-Voltage Electron Microscopy, which was held in Antwerp, September 1–3 1980. Leiden: Seventh European Congress on Electron Microscopy Foundation, 1980.

Budapest, 1984. *Electron Microscopy 1984. Proceedings of the Eighth European Congress on Electron Microscopy, Budapest* August 13–18, 1984 (Csanády, Á., Röhlich, P., & Szabó, D., Eds.). Budapest: Programme Committee of the Eighth European Congress on Electron Microscopy, 3 vols.

York, 1988. *Proceedings of the Ninth European Congress on Electron Microscopy, York,* September 4–9, 1988 (Goodhew, P. J., & Dickinson, H. G., Eds.). Bristol and Philadelphia: Institute of Physics, 1988, Conference Series 93, 3 vols.

Granada, 1992. *Electron Microscopy 92. Proceedings of the 10th European Congress on Electron Microscopy, Granada,* September 7–11, 1992 [Ríos, A., Arias, J. M., Megías-Megías, L. & López-Galindo, A., Eds. (Vol. I), López-Galindo, A. & Rodríguez-García, M. I., Eds. (Vol. II), and Megías-Megías, L., Rodríguez-García, M. I., Ríos, A. & Arias, J. M., Eds. (Vol. III)]. Granada: Secretariado de Publicaciones de la Universidad de Granada, 3 vols.

Dublin, 1996. *Electron Microscopy 1996. Proceedings of the 11th European Conference on Electron Microscopy, Dublin,* August 26–30, 1996, distributed on a defective CD-ROM. Subsequently published in book form by the Committee of European Societies of Microscopy (CESM), Brussels 1998. 3 vols.

Brno, 2000. *Electron Microscopy 2000. Proceedings of the 12th European Conference on Electron Microscopy, Brno,* July 9–14, 2000. [Frank, L., & Čiampor, F. (General Eds.); Vol. I, Biological Sciences, Čech S., & Janisch, R. (Eds.); Vol. II, Physical Sciences, Gemperlová, J., & Vávra, I. (Eds.); Vol. III, Instrumentation and Methodology, Tománek, P., & Kolařík, R. (Eds.); Vol. IV, Supplement, Frank, L., & Čiampor, F. (Eds.); Vols. I–III also distributed on CD-ROM]. Brno: Czechoslovak Society of Electron Microscopy.

Antwerp, 2004. *Proceedings of European Microscopy Congress, Antwerp,* August 23–27, 2004. [Schryvers, D., Timmermans, J.-P. & Pirard, E. (General Eds.); Biological Sciences, Verbelen, J.-P., & Wisse, E. (Eds.); Materials Sciences, Van Tendeloo, G., & Van Haesendonck, C. (Eds.); Instrumentation and Methodology, Van Dyck, D., & Van Oostveldt, P. (Eds.)]. Liège: Belgian Society for Microscopy.

Aachen, 2008. *Proceedings of EMC 2008, 14th European Microscopy Congress, Aachen,* September 1–5, 2008. Vol. 1, Instrumentation and Methods (Luysberg, M., & Tillmann, K., Eds.); Vol. 2, Materials Science (Richter, S., & Schwedt, A., Eds.); Vol. 3, Life Science (Aretz, A., Hermanns–Sachweh, B., & Mayer, J., Eds.). Berlin: Springer.

Manchester, 2012. *Proceedings of EMC2012, 15th European Microscopy Congress, Manchester,* September 16–21, 2012. Vol. 1, Physical Sciences: Applications (Stokes, D. J., & Rainforth, W. M., Eds.); Vol. 2, Physical Sciences: Tools and Techniques (Stokes, D. J., & Hutchison, J. L., Eds.); Vol. 3, Life Sciences (Stokes, D. J., O'Toole, P. J., & Wilson, T., Eds.). Oxford, UK: Royal Microscopical Society.

Lyon, 2016: August 28–September 2, 2016.

## 3.2 The Asia-Pacific Region (APEM, APMC)

In 1956, the first Asia–Pacific Conference on Electron Microscopy was held in Tokyo. The Asia–Pacific regional conferences are held under the general supervision of the Committee of Asia-Pacific Societies of Electron Microscopy (CAPSEM), now Committee of Asia-Pacific Societies of Microscopy (CAPSM). For current details of member countries, see www.ukm.my/capsm/index.htm or the IFSM website (http://ifsm.info).

Tokyo, 1956. *Electron Microscopy. Proceedings of the First Regional Conference in Asia and Oceania, Tokyo,* October 23–27, 1956. Tokyo: Electrotechnical Laboratory, 1957.

Calcutta, 1965. *Proceedings of the Second Regional Conference on Electron Microscopy in Far East and Oceania, Calcutta,* February 2–6, 1965. Calcutta: Electron Microscopy Society of India.

Singapore, 1984. *Conference Proceedings of 3rd Asia Pacific Conference on Electron Microscopy, Singapore,* August 29–September 3, 1984 (Chung Mui Fatt, Ed.). Singapore: Applied Research Corporation.

Bangkok, 1988. *Electron Microscopy 1988. Proceedings of the IVth Asia-Pacific Conference and Workshop on Electron Microscopy, Bangkok,* July 26–August 4, 1988 (Mangclaviraj, V., Banchorndhevakul, W., & Ingkaninun, P., Eds.). Bangkok: Electron Microscopy Society of Thailand.

Beijing, 1992. *Electron Microscopy I and II. 5th Asia-Pacific Electron Microscopy Conference, Beijing,* August 2–6, (Kuo, K. H., & Zhai, Z. H., Eds.). Singapore: World Scientific, 1992, 2 vols. See also *Ultramicroscopy, 48*(4), 367–490 (1993).

Hong Kong, 1996. *Proceedings of the 6th Asia-Pacific Conference on Electron Microscopy, Hong Kong,* July 1–5, 1996 (Barber, D., Chan, P. Y., Chew, E. C., Dixon, J. S., & Lai, J. K. L., Eds.). Kowloon, Hong Kong: Chinetek Promotion.

Singapore, 2000. *Proceedings of the 7th Asia-Pacific Conference on Electron Microscopy, Singapore International Convention & Exhibition Centre, Suntec City, Singapore,* June 26–30, 2000 (two volumes and CD-ROM; Yong, Y.T., Tang, C., Leong, M., Ng, C., & Netto, P., Eds.). Singapore: 7th APEM Committee.

Kanazawa, 2004. *Proceedings of 8th Asia-Pacific Conference on Electron Microscopy (8APEM), Kanazawa, Ishikawa Prefecture,* June 7–11, 2004. Full proceedings on CD-ROM. Tokyo: Japanese Society of Microscopy.

Jeju, 2008. *Proceedings of the Ninth Asia-Pacific Microscopy Conference (APMC9) Jeju, Korea,* November 2–7, 2008. Lee, H.-c., Kim, D. H., Kim, Y.-w., Rhyu, I. L., & Jeong, H.-t. (Eds.). *Korean Journal of Microscopy 38*(4) (2008), Supplement on CD-ROM only.

Perth, 2012. *Proceedings of the Tenth Asia-Pacific Microscopy Conference (APMC-10) Perth, Australia,* February 5–9, 2012 (Griffin, B., Faraone, L., & Martyniuk, M., Eds.). Held in conjunction with the 2012 International Conference on Nanoscience and Nanotechnology (ICONN2012) and the 22nd Australian Conference on Microscopy and Microanalysis (ACMM22).

Phuket, 2016. *11th Asia-Pacific Microscopy Conference (APMC-11) Phuket, Thailand,* May 23–27, 2016. Held in conjunction with the 33rd Annual Conference of the Microscopy Society of Thailand (MST-33) and the 39th Annual Conference of the Anatomy Association of Thailand (AAT-39).

## 3.3 The United States of America

In 1971, the Electron Microscopy Society of America (EMSA) began issuing a bulletin to its members; in 1993, the title of this publication changed from *EMSA Bulletin* to *MSA Bulletin,* but the latter was discontinued with the creation of the *Journal of the Microscopy Society of America* in 1995. This in turn became the journal *Microscopy and Microanalysis* in 1997; the volume numbering was continued without interruption, so the present title begins with volume 3. Since 2002, the proceedings have appeared on CD-ROM, while a small selection of these papers was printed in a supplement to *Microscopy and Microanalysis.* In 2010, the printed part was discontinued, and the CD-ROM is now the only record; however, Cambridge University Press offers a printed version of the CD-ROM on a print-on-demand basis, and a few libraries routinely carry this. In 1997, the *MSA Bulletin* was resuscitated with no break in volume numbering; the 1997 issues start with volume 23. Many other relevant publications are listed in Hawkes (2003).

EMSA 1. [First National Conference on the Electron Microscope] Hotel Sherman, Chicago, November 27–28, 1942; a transcript of the proceedings is reproduced in *EMSA and Its People, the First Fifty Years* by S. P. Newberry (Electron Microscopy Society of America, 1992).

EMSA 2. La Salle Hotel, Chicago, November 16–18, 1944; *Journal of Applied Physics, 16* (1945), 263–266.

EMSA 3. Frick Chemical Laboratory, Princeton University, Princeton, NJ, November 30–December 1, 1945; *Journal of Applied Physics, 17* (1946), 66–68.

EMSA 4. Mellon Institute, Pittsburgh, December 5–7, 1946; *Journal of Applied Physics, 18* (1947), 269–273.

EMSA 5. Franklin Institute, Philadelphia, December 11–13, 1947; *Journal of Applied Physics, 19* (1948), 118–126. See also *Analytical Chemistry, 20* (1948), 90–92.

EMSA 6. E.F. Burton Memorial Meeting, University of Toronto, 9–11 September 1948; *Journal of Applied Physics, 19* (1948) 1186–1192. See also *Analytical Chemistry, 20* (1948) 990–993.

EMSA 7. National Bureau of Standards, Washington, DC, October 6–8, 1949; *Journal of Applied Physics, 21* (1950), 66–72. See also *Analytical Chemistry, 21* (1949), 1434–1437.

EMSA 8. Statler Hotel, Detroit, September 14–16, 1950; *Journal of Applied Physics, 22* (1951), 110–116.

Washington, DC, 1951. *Electron Physics. Proceedings of the NBS Semicentennial Symposium on Electron Physics,* Washington, DC, November 5–7, 1951. Issued as *National Bureau of Standards Circular, 527* (1954). See also *Analytical Chemistry, 23* (1951), 1885–1887.

EMSA 9. Franklin Institute, Philadelphia, October 8–10, 1951; *Journal of Applied Physics, 23* (1952), 156–164. See also *Analytical Chemistry, 23* (1951), 1887–1890.

EMSA 10. Hotel Statler, Cleveland, November 6–8, 1952; *Journal of Applied Physics, 24* (1953) 111–118. See also *Analytical Chemistry, 24* (1952), 1865–1868.

EMSA 11. Pocono Manor Inn, Pocono Manor, PA, November 5–7, 1953; *Journal of Applied Physics, 24* (1953), 1414–1426. See also *Analytical Chemistry, 26* (1954), 437–438.

EMSA 12. Moraine-on-the-Lake Hotel, Highland Park, IL, October 14–16, 1954; *Journal of Applied Physics, 25* (1954), 1453–1468.

EMSA 13. Pennsylvania State University, University Park, PA, October 27–29, 1955; *Journal of Applied Physics, 26* (1955), 1391–1398.

EMSA 14. University of Wisconsin, Madison, WI, September 10–12, 1956; *Journal of Applied Physics, 27* (1956), 1389–1398.

EMSA 15. Massachusetts Institute of Technology, Cambridge, MA, September 9–11, 1957; *Journal of Applied Physics, 28* (1957), 1368–1386.

EMSA 16. Santa Monica Civic Auditorium, Santa Monica, CA, August 7–9, 1958; *Journal of Applied Physics, 29* (1958), 1615–1626.

EMSA 17. Ohio State University, Columbus, OH, September 9–12, 1959; *Journal of Applied Physics, 30* (1959), 2024–2042.

EMSA 18. Marquette University, Milwaukee, WI, August 29–31, 1960; *Journal of Applied Physics, 31* (1960), 1831–1848.

EMSA 19. Pittsburgh Hilton Hotel, Pittsburgh, 23–26 August 1961; *Journal of Applied Physics, 32* (1961), 1626–1646.

ICEM–5, Philadelphia, August 29–September 5, 1962.

EMSA 21. Denver, August 28–31, 1963; *Journal of Applied Physics, 34* (1963), 2502–2534.

EMSA 22. Detroit, October 13–16, 1964; *Journal of Applied Physics, 35* (1964), 3074–3102.

EMSA 23. New York, August 25–28, 1965; *Journal of Applied Physics, 36* (1965), 2603–2632.

EMSA 24. San Francisco Hilton, San Francisco, August 22–25, 1966; *Journal of Applied Physics, 37* (1966), 3919–3952.

EMSA 25. *Proceedings of the 25th Anniversary Meeting Electron Microscopy Society of America,* Chicago, August 29–September 1, 1967 (Arceneaux, C. J., Ed.); Baton Rouge: Claitor (1967).

EMSA 26. *Proceedings of the 26th Annual Meeting Electron Microscopy Society of America,* New Orleans, September 16–19, 1968 (Arceneaux, C. J., Ed.); Baton Rouge: Claitor (1968).

EMSA 27. *Proceedings of the 27th Annual Meeting Electron Microscopy Society of America,* St Paul, August 26–29, 1969 (Arceneaux, C. J., Ed.); Baton Rouge: Claitor (1969).

EMSA 28. *Proceedings of the 28th Annual Meeting Electron Microscopy Society of America,* Houston TX, 5–8 October 1970 (Arceneaux, C. J., Ed.); Baton Rouge: Claitor (1970).

EMSA 29. *Proceedings of the 29th Annual Meeting Electron Microscopy Society of America,* Boston, August 9–13, 1971 (Arceneaux, C. J., Ed.); Baton Rouge: Claitor (1971).

EMSA 30. *Proceedings of the 30th Annual Meeting Electron Microscopy Society of America and First Pacific Regional Conference on Electron Microscopy,* Los Angeles, August 14–17, 1972 (Arceneaux, C. J., Ed.); Baton Rouge: Claitor (1972).

EMSA 31. *Proceedings of the 31st Annual Meeting Electron Microscopy Society of America,* New Orleans, August 14–17, 1973 (Arceneaux, C. J., Ed.); Baton Rouge: Claitor (1973).

EMSA 32. *Proceedings of the 32nd Annual Meeting Electron Microscopy Society of America,* St. Louis, August 13–15, 1974 (Arceneaux, C. J., & G. W. Bailey, G. W., Eds.); Baton Rouge: Claitor (1974).

EMSA 33. *Proceedings of the 33rd Annual Meeting Electron Microscopy Society of America,* Las Vegas, August 11–15, 1975 (Bailey, G. W., & Arceneaux, C. J., Eds.); Baton Rouge: Claitor (1975).

EMSA 34. *Proceedings of the 34th Annual Meeting Electron Microscopy Society of America,* Miami Beach, August 9–13, 1976 (Bailey, G. W., Ed.); Baton Rouge: Claitor (1976).

EMSA 35. *Proceedings of the 35th Annual Meeting Electron Microscopy Society of America,* Boston, August 22–26, 1977 (Bailey, G. W., Ed.); Baton Rouge: Claitor (1977).

ICEM-9, Toronto, August 1–9, 1978 (incorporates EMSA 36).

EMSA 37. *Proceedings of the 37th Annual Meeting Electron Microscopy Society of America,* San Antonio, TX, August 13–17, 1979 (Bailey, G. W., Ed.); Baton Rouge: Claitor (1979).

EMSA 38. *Proceedings of the 38th Annual Meeting Electron Microscopy Society of America,* San Francisco, August 4–8, 1980 (Bailey, G. W., Ed.); Baton Rouge: Claitor (1980).

EMSA 39. *Proceedings of the 39th Annual Meeting Electron Microscopy Society of America,* Atlanta, August 10–14, 1981 (Bailey, G. W., Ed.); Baton Rouge: Claitor (1981).

EMSA 40. *Proceedings of the 40th Annual Meeting Electron Microscopy Society of America,* Washington, DC, August 9–13, 1982 (Bailey, G. W., Ed.); Baton Rouge: Claitor (1982).

EMSA 41. *Proceedings of 41st Annual Meeting Electron Microscopy Society of America,* Phoenix, August 8–12, 1983 (Bailey, G. W., Ed.); San Francisco: San Francisco Press (1983). See also *Ultramicroscopy 13* (1984), 1–183.

EMSA 42. *Proceedings of 42nd Annual Meeting Electron Microscopy Society of America jointly with Microscopical Society of Canada, Eleventh Annual Meeting,* Detroit, August 13–17, 1984 (Bailey, G. W., Ed.); San Francisco: San Francisco Press (1984).

EMSA 43. *Proceedings of 43rd Annual Meeting Electron Microscopy Society of America,* Louisville, KY, August 5–9, 1985 (Bailey, G. W., Ed.); San Francisco: San Francisco Press (1985). See S. Basu and J. R. Millette (Eds), *Electron Microscopy in Forensic, Occupational and Environmental Health Sciences.* New York & London: Plenum (1986).

EMSA 44. *Proceedings of 44th Annual Meeting Electron Microscopy Society of America,* Albuquerque, NM, August 10–15, 1986 (Bailey, G. W., Ed.); San Francisco: San Francisco Press (1986).

EMSA 45. *Proceedings of 45th Annual Meeting Electron Microscopy Society of America,* Baltimore, August 2–7, 1987 (Bailey, G. W., Ed.); San Francisco: San Francisco Press (1987).

EMSA 46. *Proceedings of 46th Annual Meeting Electron Microscopy Society of America jointly with Microscopical Society of Canada, Fifteenth Annual Meeting,* Milwaukee, August 7–12, 1988 (Bailey, G. W., Ed.); San Francisco: San Francisco Press (1988).

EMSA 47. *Proceedings of 47th Annual Meeting Electron Microscopy Society of America,* San Antonio, TX, August 6–11, 1989 (Bailey, G. W., Ed.); San Francisco: San Francisco Press (1989).

ICEM-XII, Seattle, August 12–18, 1990 (incorporates EMSA 48).

EMSA 49. *Proceedings of 49th Annual Meeting Electron Microscopy Society of America,* San Jose, CA, August 4–9, 1991 (Bailey, G. W., & Hall, E. L.,

Eds..); San Francisco: San Francisco Press (1991). See also *Ultramicroscopy* 47(1–3), 1–306 (1992).

EMSA 50. *Proceedings of 50th Annual Meeting Electron Microscopy Society of America, 27th Annual Meeting Microbeam Analysis Society, Nineteenth Annual Meeting Microscopical Society of Canada/Société de Microscopie du Canada*, Boston, August 16–21, 1992 (Bailey, G. W., Bentley, J., & Small, J. A., Eds.); San Francisco: San Francisco Press (1992), 2 vols.

MSA 51. *Proceedings of 51st Annual Meeting Microscopy Society of America*, Cincinnati, OH, August 1–6, 1993 (Bailey, G. W., & Rieder, C. L., Eds.); San Francisco: San Francisco Press (1993).

MSA 52. *Proceedings of 52nd Annual Meeting Microscopy Society of America, 29th Annual Meeting Microbeam Analysis Society*, New Orleans, July 31–August 5, 1994 (Bailey, G. W., & Garratt-Reed, A. J., Eds.); San Francisco: San Francisco Press (1994).

MSA 53. *Microscopy and Microanalysis 1995,* 53rd Annual Meeting Microscopy Society of America, Kansas City, KS, August 13–16, 1995; *Journal of the Microscopy Society of America Proceedings* (Bailey, G. W., Ellisman, M. H., Henniger, R. A., & Zaluzec, N. J., Eds.; New York and Boston: Jones and Begell (1995).

MSA 54. *Proceedings of Microscopy and Microanalysis 1996.* 54th Annual Meeting Microscopy Society of America, Twenty-third Annual Meeting Microscopical Society of Canada/Société de Microscopie du Canada, 30th Annual Meeting Microbeam Analysis Society, Minneapolis, MN, August 11–15, 1996 (Bailey, G. W., Corbett, J. M., Dimlich, R. V. W., Michael, J. R., & Zaluzec, N. J., Eds.); San Francisco: San Francisco Press (1996).

MSA 55. *Proceedings of Microscopy and Microanalysis 1997.* 55th Annual Meeting Microscopy Society of America, 31st Annual Meeting Microbeam Analysis Society, 48th Annual Meeting Histochemical Society, Cleveland, OH, August 10–14, 1997; *Microscopy and Microanalysis, 3* (1997), Supplement 2 (Bailey, G. W., Dimlich, R. V. W., Alexander, K. B., McCarthy, J. J., & Pretlow, T. P., Eds.); New York: Springer (1997).

MSA 56. *Proceedings of Microscopy and Microanalysis 1998.* 56th Annual Meeting Microscopy Society of America, 32nd Annual Meeting Microbeam Analysis Society, Atlanta, July 12–16, 1998; *Microscopy and Microanalysis, 4* (1998), Supplement 2 (Bailey, G. W., Alexander, K.B., Jerome, W.G., Bond, M.G. & McCarthy, J.J., Eds.); New York: Springer (1998).

MSA 57. *Proceedings of Microscopy and Microanalysis 1999.* 57th Annual Meeting Microscopy Society of America, 33rd Annual Meeting Microbeam Analysis Society, Portland, OR, August 1–5, 1999; *Microscopy and*

*Microanalysis, 5* (1999), Supplement 2 (Bailey, G. W., Jerome, W. G., McKernan, S., Mansfield, J. F., & Price, R. L., Eds.); New York: Springer (1999).

MSA 58. *Proceedings of Microscopy and Microanalysis 2000.* 58th Annual Meeting Microscopy Society of America, 27th Annual Meeting Microscopical Society of Canada/Société de Microscopie du Canada, 34th Annual Meeting Microbeam Analysis Society, Philadelphia, August 13–17, 2000; *Microscopy and Microanalysis, 6* (2000), Supplement 2 (Bailey, G. W., McKernan, S., Price, R. L., Walck, S. D., Charest, P.-M., & Gauvin, R., Eds.); New York: Springer (2000).

MSA 59. *Proceedings of Microscopy and Microanalysis 2001.* 59th Annual Meeting Microscopy Society of America, 35th Annual Meeting Microbeam Analysis Society, Long Beach, CA, August 5–9, 2001; *Microscopy and Microanalysis, 7* (2001), Supplement 2 (Bailey, G. W., Price, R. L., Voelkl, E., & Musselman, I. H., Eds.); New York: Springer (2001).

MSA 60. *Proceedings of Microscopy and Microanalysis 2002.* 60th Annual Meeting Microscopy Society of America, 29th Annual Meeting Microscopical Society of Canada/Société de Microscopie du Canada, 36th Annual Meeting Microbeam Analysis Society, 35th Meeting International Metallographic Society, Quebec City, PQ, Canada, August 4–8, 2002. *Microscopy and Microanalysis, 8* (2002), Supplement 2 (Voelkl, E., Piston, D., Gauvin, R., Lockley, A. J., Bailey, G. W., & McKernan, S., Eds.); New York: Cambridge University Press (2002). The printed proceedings contain only the invited papers; all the others are distributed on a CD-ROM.

MSA 61. 61st Annual Meeting Microscopy Society of America, 7th Interamerican Congress on Electron Microscopy, 37th Annual Meeting Microbeam Analysis Society, 36th Meeting International Metallographic Society, San Antonio, TX, August 3–7, 2003. *Microscopy and Microanalysis, 9* (2003), Supplement 2 (Piston, D., Bruley, J. T., Anderson, I. M., Kotula, P., Solorzano, G., Lockley, A., & McKernan, S., Eds.).

MSA 62. 62nd Annual Meeting Microscopy Society of America, 38th Annual Meeting Microbeam Analysis Society, 37th Meeting International Metallographic Society Savannah, GA, August 1–5, 2004. *Microscopy and Microanalysis, 10* (2004) Supplement 2 (Anderson, I. M., Price, R., Hall, E., Clark, E., & McKernan, S., Eds.).

MSA 63. 63rd Annual Meeting Microscopy Society of America, 39th Annual Meeting Microbeam Analysis Society, Honolulu, HI, July 31–August 4, 2005. *Microscopy and Microanalysis, 11* (2005), Supplement 2 (Price, R., Kotula, P., Marko, M., Scott, J. H., Vander Voort, G. F., Manilova, E., Mah Lee Ng, M., Smith, K., Griffin, B., Smith, P., & McKernan, S., Eds.).

MSA 64. 64th Annual Meeting Microscopy Society of America, 40th Annual Meeting Microbeam Analysis Society, 40th Meeting International Metallographic Society, 33rd Meeting Microscopy Society of Canada, Chicago, July 30–August 3, 2006. *Microscopy and Microanalysis, 12* (2006), Supplement 2 (Kotula, P., Marko, M., Scott, J.-H., Gauvin, R., Beniac, D., Lucas, G., McKernan, S., & Shields, J., Eds.).

MSA 65. 65th Annual Meeting Microscopy Society of America, 41st Annual Meeting Microbeam Analysis Society, 40th Meeting International Metallographic Society, Fort Lauderdale, FL, August 5–9, 2007. *Microscopy and Microanalysis, 13* (2007), Supplement 2 (Marko, M., Scott, J. H., Vicenzi, E., Dekanich, S., Frafjord, J., Kotula, P., McKernan, S., & Shields, J., Eds.).

MSA 66. 66th Annual Meeting Microscopy Society of America, 42nd Annual Meeting Microbeam Analysis Society, 41st Meeting International Metallographic Society, Albuquerque, NM, August 3–7, 2008. *Microscopy and Microanalysis, 14* (2008), Supplement 2 (Marko, M., McKernan, S., Shields, J., Scott, J.-H., Kotula, P., Anderson, I. & Woodward, J., Eds.).

MSA 67. 67th Annual Meeting Microscopy Society of America, 43rd Annual Meeting Microbeam Analysis Society, 42nd Meeting International Metallographic Society, Richmond, VA, July 26–30, 2009. *Microscopy and Microanalysis, 15* (2009) Supplement 2 (Brewer, L. N., McKernan, S., Shields, J. P., Schmidt, F. E., Woodward, J. H., & Zaluzec, N., Eds.).

MSA 68. 68th Annual Meeting Microscopy Society of America, 44th Annual Meeting Microbeam Analysis Society, 43rd Meeting International Metallographic Society, 37th Meeting Microscopy Society of Canada, Portland, OR, August 1–5, 2010. *Microscopy and Microanalysis, 16* (2010), Supplement 2 [print on demand only from now on] (Shields, J., McKernan, S., Mansfield, J., Foran, B., Frafjord, J., Beniac, D. & O'Neil, D., Eds.).

MSA 69. 69th Annual Meeting Microscopy Society of America, 45th Annual Meeting Microbeam Analysis Society, 44th Meeting International Metallographic Society, Nashville, TN, August 7–11, 2011. *Microscopy and Microanalysis, 17* (2011), Supplement 2 (Shields, J., McKernan, S., Giovannucci, D., Watanabe, M., & Susan, D., Eds.).

MSA 70. 70th Annual Meeting Microscopy Society of America, 46th Annual Meeting Microbeam Analysis Society, 45th Meeting International Metallographic Society, Phoenix, July 29–August 2, 2012. *Microscopy and Microanalysis, 18* (2012), Supplement 2 (Shields, J., McKernan, S., Brewer, L., Ruiz, T., & Turnquist, D., Eds.).

MSA 71. 71st Annual Meeting Microscopy Society of America, 47th Annual Meeting Microbeam Analysis Society, 46th Meeting International

Metallographic Society, Indianapolis, IN, August 4–8, 2013. *Microscopy and Microanalysis, 19* (2013), Supplement 2.

MSA 72. 72nd Annual Meeting Microscopy Society of America, 48th Annual Meeting Microbeam Analysis Society, 47th Meeting International Metallographic Society, 41st Meeting Microscopy Society of Canada, 6th Meeting International Union of Microbeam Analysis Societies, Hartford, CT, August 3–7, 2014. *Microscopy and Microanalysis, 20* (2014), Supplement 2.

MSA 73. Portland, OR, August 2–6, 2015
MSA 74. Columbus, OH, July 24–28, 2016
MSA 75. St Louis, MO, July 23–27, 2017
MSA 76. Baltimore, MD, August 5–9, 2018
MSA 77. Portland, OR, August 4–8, 2019

## 3.4 South America

The first society grouping the electron microscopists of South America was the Sociedad Latinoamericana de Microscopía Electrónica (SLAME), founded in 1972 by Argentina, Brazil, Chile, Colombia, and Venezuela. The *Revista de Microscopía Electrónica*, which has changed names several times over the years (*Revista de Microscopía Electrónica y Biología Celular*, *Microscopía Electrónica y Biología Celular*, and *Biocell*), was the official organ of the society, and later of other learned societies in this field as well. Today, however, a new association has been formed, the Committee of Inter-American Societies of Electron Microscopy (CIASEM), the members of which are listed on the CIASEM website (www.ciasem.com). The various regional congresses organized by SLAME and CIASEM are listed here. Proceedings are usually published online as supplements to the open access journal *Acta Microscópica*.

First Latin-American Congress of Electron Microscopy, Maracaibo (Venezuela), May 26–30, 1972. *Revista de Microscopía Electrónica 1* (1972) Nos 1 (Resumenes) and 2 (Simposios).

Second Latin-American Congress for Electron Microscopy and IV Colóquio Brazileiro de Microscopia Electrônica, Faculdade de Medicina, Ribeirão Preto SP (Brazil), December 1–5, 1974 [Proceedings not published in a serial].

Third Latin-American Congress for Electron Microscopy, Santiago (Chile), November 22–26, 1976. *Revista de Microscopía Electrónica 3*(1) (1976) (Resumenes) and *4*(1) (1977) (Simposios). Cf. M. Ipohorski, Electron microscopy in Latin America. *Ultramicroscopy 3*, 97 (1978).

Fourth Latin-American Congress for Electron Microscopy, Mendoza (Argentina) October 12–18, 1978. *Revista de Microscopía Electrónica 5*(1) (1978), and *Revista de Microscopía Electrónica y Biología Celular 6*(1–2) (1979).

Fifth Latin-American Congress for Electron Microscopy, Bogotá (Columbia), November 17–20, 1981 [Proceedings not published in a serial].

Sixth Latin-American Congress for Electron Microscopy, Maracaibo (Venezuela), December 2–8, 1984. *Revista de Microscopía Electrónica y Biología Celular 8*(2) (1984), and Supplement.

Seventh Latin-American Congress for Electron Microscopy, Barcelona (Spain), March 16–18, 1987 [Proceedings not published in a serial].

Eighth Latin-American Congress for Electron Microscopy, La Habana (Cuba), May 2–5, 1989 [Proceedings not published in a serial].

## NEW SERIES

Memorias del 1er. Congreso Atlántico de Microscopía Electrónica/ Proceedings of the First Atlantic Congress of Electron Microscopy, Facultad de Ciencias, Universidad de los Andes, Mérida (Venezuela), May 25–29, 1992, Valiente, E. (Ed.) (Editorial Venezolana, Mérida, 1992). Held jointly with the V Jornadas Venezolanas de Microscopía Electrónica.

Second Interamerican Conference on Electron Microscopy, Cancún (México), September 26–October 1, 1993.

Third Interamerican Conference on Electron Microscopy and XVth Meeting of the Brazilian Society of Electron Microscopy, Caxambú MG (Brazil), September 2–6, 1995. *Acta Microscópica 4* (1995), Supplements A (Biological Science) and B (Materials Science).

Fourth Interamerican Congress on Electron Microscopy and II Ecuadorian Congress on Electron Microscopy related to Medical, Biological and Materials Sciences, Guayaquil (Ecuador), September 23–26, 1997. *Acta Microscópica 6*, 39–57 (1997) [posters].

Fifth Interamerican Congress on Electron Microscopy, Margarita Island (Venezuela), October 24–28, 1999. Full proceedings were produced as a CD-ROM [Carrasquero, E., Castañeda, G., & Ramos, E. (Eds.); preface by A. Castellano].

Sixth Interamerican Congress on Electron Microscopy, Veracruz (Mexico), October 7–11, 2001. *Acta Microscópica*, October 2001, 614 pp.

Seventh Interamerican Congress on Electron Microscopy, San Antonio TX (USA), 3–7 August 2003. Held jointly with *Microscopy and Microanalysis,* 2003. *Microscopy and Microanalysis, 9* (2003), Supplememt 2 (Piston, D.,

Bruley, J., Anderson, I. M., Kotula, P., Solorzano, G., Lockley, A., & McKernan, S., Eds.).

Eighth Interamerican Congress on Electron Microscopy, La Habana (Cuba), September 25–30, 2005. Proceedings available only on CD-ROM (distributed at the conference).

Ninth Interamerican Congress on Electron Microscopy, Cusco, Peru, September 23–28, 2007. *Acta Microscópica 16*(1–2), Supplement B (2007).

Tenth Interamerican Congress on Electron Microscopy, Rosario, Argentina, October 25–28, 2009. *Acta Microscópica 18* (2009), Supplement C (Tolley,, A., Condó, A. M., Capani, F., Pelegrina, J. L., Alvarez, I., & Lovey, F. C. (Eds.).

Eleventh Interamerican Congress on Microscopy, Mérida, Yucatan, Mexico, September 25–30, 2011. *Acta Micróscopica 20* (2011) Supplement B. Gasga, J. R., Alatorre, J. A., & Flores, J. B. K. (Eds.).

Twelfth Interamerican Congress on Microscopy, Cartagena, Colombia, September 24–28, 2013. *Acta Microscópica 22* (2013) Supplement A.

Thirteenth Interamerican Congress on Microscopy, Isla Margarita, Venezuela, October 18–23, 2015. *Acta Microscópica 24* (2015).

## 4. CONGRESSES ON ELECTRON MICROSCOPY ORGANIZED BY SEVERAL NATIONAL SOCIETIES

### 4.1 Multinational Congresses on Electron Microscopy (MCEM, MCM)

The first of the Multinational Congresses on Electron Microscopy brought together the Italian, Hungarian, Czechoslovak, and Slovenian societies. For subsequent meetings, these were joined by the Austrian and Croatian societies.

MCEM-1993. Multinational Congress on Electron Microscopy, Parma, September 13–17, Proceedings issued as a supplement to *Microscopia Elettronica 14*(2).

MCEM-95. Proceedings of Multinational Conference on Electron Microscopy, Stará Lesná (High Tatra Mountains), October 16–20, 1995. Bratislava: Slovak Academic Press.

MCEM-97. Proceedings of Multinational Congress on Electron Microscopy, Portorož å (Slovenia), October 5–8, 1997. Part I: Microscopy Applications in the Life Sciences; Part II: Microscopy Applications in the Material Sciences; Part III: Microscopy Methods and Instrumentation. *Journal of Computer-assisted Microscopy*, *8*(4) (1996), and *9*(1–2) (1997).

MCEM-99. Proceedings of 4th Multinational Congress on Electron Microscopy, Veszprém (Hungary), September 5–8, 1999 (Kovács, K., Ed.). University of Veszprem, 1999.

MCEM-5. Proceedings of the 5th Multinational Congress on Electron Microscopy, Department of Biology, University of Lecce (Italy), September 20–25, 2001 (Dini, L., & Catalano, M., Eds.). Princeton, NJ: Rinton Press, 2001.

MCM-6. Proceedings of the Sixth Multinational Congress on Microscopy, Pula (Croatia), June 1–5, 2003 (Milat, O., & Ježek, D., Eds.). Zagreb: Croatian Society for Electron Microscopy, 2003.

MCM-7. Proceedings of the 7th Multinational Congress on Microscopy, Portorož, (Slovenia) June 26–30, 2005 (Čeh, M., Dražič, G., & Fidler, S., Eds.). Ljubljana: Slovene Society for Microscopy and Department for Nanostructured Materials, Jožef Stefan Institute, 2005.

MCM-8. Proceedings of 8th Multinational Congress on Microscopy, Prague (Czech Republic), June 17–21, 2007 (Nebesářova, J., & Hozák, P., Eds.). Prague: Czechoslovak Microscopy Society, 2007.

MC 2009 incorporating MCM-9. Microscopy Conference, Graz, Austria, August 30–September 4, 2009. First Joint Meeting of Dreiländertagung and Multinational Congress on Microscopy. Vol. 1, Instrumentation and Methodology (Kothleitner, G., & Leisch, M., Eds.); Vol. 2, Life Sciences (Pabst, M. A., & Zellnig, G., Eds.); Vol. 3, Materials Science (Grogger, W., Hofer, F., & Pölt, P., Eds.). Graz: Verlag der Technischen Universität, 2009.

MCM-10. Proceedings of 10th Multinational Conference on Microscopy, Urbino, September 4–9, 2011 (Falcieri, E., Ed.). Società Italiana di Scienze Microscopiche (SISM), 2011.

MCM-11. Microscopy Conference (MC 2013) Regensburg, August 25–30, 2013. Joint Meeting of Dreiländertagung and Multinational Congress on Microscopy, together with the Serbian and Turkish Microscopy Societies. Proceedings can be downloaded from www.mc2013.de. urn: nbn:de: bvb:355-epub-287343 (Rachel, R., Schröder, J., Witzgall, R., & Zweck, J., Eds.).

MCM-12. Multinational Conference on Microscopy, Eger (Hungary), August 23–29, 2015.

## 4.2 The Dreiländertagungen (Germany, Austria, Switzerland) and Related Meetings

These conferences are organized in turn by the Austrian, German, and Swiss Microscopy societies; originally designed for German-speaking microscopists, they now tend to encourage presentations in English, hence attracting wider participation.

Dreiländertagung für Elektronenmikroskopie. Konstanz, September 15–21, 1985. *Optik* (1985) Supplement 1; or *European Journal of Cell Biology* (1985) Supplement 10. See also *Beiträge zur Elektronenmikroskopische Direktabbildung von Oberflächen 18* (1985).

Dreiländertagung für Elektronenmikroskopie. Salzburg, September 10–16, 1989. *Optik 83* (1989) Supplement 4; or *European Journal of Cell Biology 49* (1989) Supplement 27.

Dreiländertagung für Elektronenmikroskopie. Zürich, September 5–11, 1993. *Optik 94* (1993) Supplement 5; or *European Journal of Cell Biology 61* (1993) Supplement 39.

Dreiländertagung für Elektronenmikroskopie. Regensburg, September 7–12, 1997. *Optik 106* (1997) Supplement 7; or *European Journal of Cell Biology 74* (1997) Supplement 45.

Dreiländertagung für Elektronenmikroskopie. Innsbruck, September 9–14, 2001. Abstracts book (168 pp) not published as a Supplement to *Optik* or *European Journal of Cell Biology*.

MC-2003, Dresden, September 7–12, 2003. *Microscopy and Microanalysis, 9* (2003), Supplement 3 (Gemming, T., Lehmann, M., Lichte, H., & Wetzig, K., Eds.).

Dreiländertagung für Elektronenmikroskopie. Microscopy Conference 2005, Paul Scherrer Institute, Davos, August 25–September 2, 2005. *Paul-Scherrer-Institute Proceedings PSI 05–01*, 2005.

MC-2007, Saarbrücken, September 2–7, 2007. *Microscopy and Microanalysis, 13* (2007) Supplement 3 (Gemming, T., Hartmann, U., Mestres, P., & Walther, P., Eds.).

Microscopy Conference (MC 2009), Graz, August 30–September 4, 2009. First Joint Meeting of Dreiländertagung & Multinational Congress on Microscopy. Vol. 1, Instrumentation and Methodology (Kothleitner, G., & Leisch, M., Eds.); Vol. 2, Life Sciences (Pabst, M. A. & Zellnig, G., Eds.); Vol. 3, Materials Science (Grogger, W., Hofer, F., & Pölt, P., Eds.). Graz: Verlag der Technischen Universität, 2009.

MC-2011, Kiel, August 28–September 2, 2011. Joint meeting of the German Society (Deutsche Gesellschaft für Elektronenmikroskopie,

DGE), the Nordic Microscopy Society (SCANDEM), and the Polish Microscopy Society (PTMi), with the participation of microscopists from Estonia, Latvia, Lithuania, and St. Petersburg, Russia. Proceedings published in 3 volumes by the German Society for Electron Microscopy and also distributed as a USB key (Jäger, W., Kaysser, W., Benecke, W., Depmeier, W., Gorb, S., Kienle, L., Mulisch, M., Häußler, D., & Lotnyk, A., Eds.).

Microscopy Conference (MC 2013) Regensburg, August 25–30, 2013. Joint Meeting of Dreiländertagung and Multinational Congress on Microscopy, together with the Serbian and Turkish Microscopy Societies. Proceedings can be downloaded from www.mc2013.de. urn:nbn:de: bvb:355-epub-287343 (Rachel, R., Schröder, J., Witzgall, R., & Zweck, J., Eds.).

Microscopy Conference (MC 2015) Georg-August-Universität, Göttingen, September 6–11, 2015.

# 5. CHARGED-PARTICLE OPTICS CONFERENCES

## 5.1 Charged-Particle Optics (CPO)

The CPO series of meetings was launched by Hermann Wollnik, in collaboration with Karl L. Brown and Peter W. Hawkes, in an attempt to bring together members of the various communities of charged-particle optics (accelerator optics, electron optics, and spectrometer optics).

Giessen, 1980. Proceedings of the First Conference on Charged Particle Optics, Giessen, September 8–11, 1980 (Wollnik, H., Ed.) *Nuclear Instruments and Methods in Physics Research, 187* (1981), 1–314.

Albuquerque, NM, 1986. Proceedings of the Second International Conference on Charged Particle Optics, Albuquerque, NM, May 19–23, 1986 (Schriber, S. O., & Taylor, L. S., Eds.) *Nuclear Instruments and Methods in Physics Research A, 258* (1987), 289–598.

Toulouse, France, 1990. Proceedings of the Third International Conference on Charged Particle Optics, Toulouse, April 24–27, 1990 (Hawkes, P. W., Ed.) *Nuclear Instruments and Methods in Physics Research A, 298* (1990), 1–508.

Tsukuba, Japan, 1994. Proceedings of the Fourth International Conference on Charged Particle Optics, Tsukuba, October 3–6, 1994 (Ura, K., Hibino, M., Komuro, M., Kurashige, M., Kurokawa, S., Matsuo, T., Okayama, S., Shimoyama, H., & Tsuno, K., Eds.) *Nuclear Instruments and Methods in Physics Research A 363* (1995), 1–496.

Delft, the Netherlands, 1998. Proceedings of the Fifth International Conference on Charged Particle Optics, Delft, April 14–17, 1998 (Kruit, P., & Amersfoort, van, P. W. Eds.). *Nuclear Instruments and Methods in Physics Research A, 427* (1999), 1–422.

College Park, MD, 2002. Proceedings of the Sixth International Conference on Charged Particle Optics, Marriott Hotel, Greenbelt, MD, October 21–25, 2002 (Dragt, A. & Orloff, J., Eds.). *Nuclear Instruments and Methods in Physics Research A, 519* (2004), 1–487.

Cambridge, UK, 2006. Proceedings of the Seventh International Conference on Charged Particle Optics, Trinity College, Cambridge, July 24–28, 2006 (Munro, E., & Rouse, J., Eds.). *Physics Procedia 1* (2008), 1–572.

Singapore, 2010. Proceedings of the Eighth International Conference on Charged Particle Optics, Suntec Convention Centre, Singapore, July 12–16, 2010 (Khursheed, A., Hawkes, P. W., & Osterberg, M. B., Eds.). *Nuclear Instruments and Methods in Physics Research A, 645* (2011), 1–354.

Brno, Czech Republic, 2014. Proceedings of the Ninth International Conference on Charged Particle Optics, Brno, August 31–September 5, 2014. (Frank, L., Hawkes, P.W., & Radlička, T., Eds.). Microscopy and Microanalysis, *21* (2015) Suppl.

Key West 2018

## 5.2 Problems of Theoretical and Applied Electron Optics [Problemyi Teoreticheskoi i Prikladnoi Elektronnoi Optiki][1]

Proceedings of the First All-Russia Seminar, Scientific Research Institute for Electron and Ion Optics, Moscow 1996. *Prikladnaya Fizika* (1996), No. 3.

Proceedings of the Second All-Russia Seminar, Scientific Research Institute for Electron and Ion Optics, Moscow, April 25, 1997. *Prikladnaya Fizika* (1997), Nos 2–3.

Proceedings of the Third All-Russia Seminar, Scientific Research Institute for Electron and Ion Optics, Moscow, March 31–April 2, 1998. *Prikladnaya Fizika* (1998), Nos 2 and 3/4.

Proceedings of the Fourth All-Russia Seminar, Scientific Research Institute for Electron and Ion Optics, Moscow, October 21–22, 1999. *Prikladnaya Fizika* (2000), Nos 2 and 3; *Proceedings of SPIE 4187* (2000), Filachev, A. M., & Gaidoukova, I. S. (Eds.).

---

[1] For *Prikladnaya Fizika*, see applphys.vimi.ru.

Proceedings of the Fifth All-Russia Seminar, Scientific Research Institute for Electron and Ion Optics, Moscow, November 14–15, 2001. *Prikladnaya Fizika* (2002) No. 3; *Proceedings of SPIE* (2003) *5025*, Filachev, A. M. (Ed.).

Proceedings of the Sixth All-Russia Seminar, Scientific Research Institute for Electron and Ion Optics, Moscow, May 28–30, 2003. *Prikladnaya Fizika* (2004), No. 1; and *Proceedings of SPIE 5398* (2004), Filachev, A. M., & Gaidoukova, I. S. (Eds.).

Proceedings of the Seventh All-Russia Seminar, Scientific Research Institute for Electron and Ion Optics, Moscow, May 25–27, 2005. *Prikladnaya Fizika* (2006), No. 3; and *Proceedings of SPIE* 6278 (2004), Filachev, A. M., & Gaidoukova, I. S. (Eds.).

Proceedings of the Eighth All-Russia Seminar, Scientific Research Institute for Electron and Ion Optics, Moscow, May 29–31, 2007. *Prikladnaya Fizika* (2008), No. 2; and *Proceedings of SPIE* 7121 (2008), Filachev, A. M., & Gaidoukova, I. S. (Eds.).

Proceedings of the Ninth All-Russia Seminar, Scientific Research Institute for Electron and Ion Optics, Moscow, May 28–31, 2009. *Prikladnaya Fizika* (2010), No. 3, 31–115, Filachev, A. M. (Eds.) & Gaidoukova, I. S. (Eds.).

Proceedings of the Tenth All-Russia Seminar, Scientific Research Institute for Electron and Ion Optics, Moscow, May 24–26, 2011. *Prikladnaya Fizika* (2012), No. 2, Dirochka, A. L., & Filachev, A. M. (Eds.).

Proceedings of the Eleventh All-Russia Seminar, Scientific Research Institute for Electron and Ion Optics, Moscow, May 28–30, 2013. *Uspekhi Prikladnoi Fiziki* 1 (2013), No. 5, 571–600.

## 5.3 Recent Trends in Charged-particle Optics and Surface Physics Instrumentation (Skalský Dvůr)

1989. First Seminar, Brno, Czechoslovakia, September 4–6, 1989.

1990. Second Seminar, Brno, Czechoslovakia, September 27–29, 1990.

1992. Third Seminar, Skalský Dvůr, Czechoslovakia, June 15–19, 1992.

1994. Fourth Seminar, Skalský Dvůr, Czech Republic, September 5–9, 1994.

1996. Fifth Seminar, Skalský Dvůr, Czech Republic, June 24–28, 1996. Proceedings edited by I. Müllerová & L. Frank (92 pp).

1998. Sixth Seminar, Skalský Dvůr, Czech Republic, June 29–July 3, 1998. Proceedings edited by I. Müllerová & L. Frank (84 pp). Published by the CSEM (Brno, Czech Republic, 1998).

2000. 7th Seminar, Skalský Dvůr, Czech Republic, July 15–19, 2000.

2002. 8th Seminar, Skalský Dvůr, Czech Republic, July 8–12, 2002. Proceedings (96 pp + Supplement, 6 pp) edited by L. Frank. Published by the CSMS (Brno, Czech Republic, 2002).

2004. 9th Seminar, Skalský Dvůr, Czech Republic, July 12–16, 2004. Proceedings edited by I. Müllerová. Published by the CSMS (Brno, Czech Republic, 2004).

2006. 10th Seminar, Skalský Dvůr, Czech Republic, May 22–26, 2006. Proceedings edited by I. Müllerová. Published by the CSMS (Brno, Czech Republic, 2006).

2008. 11th Seminar, Skalský Dvůr, Czech Republic, July 14–18, 2008. Proceedings edited by F. Mika (Published by the CSMS (Brno, Czech Republic, 2008).

2010. 12th Seminar, Skalský Dvůr, Czech Republic, May 31–June 4, 2010. Proceedings edited by F. Mika. Published by the CSMS (Brno, Czech Republic, 2010).

2012. 13th Seminar, Skalský Dvůr, Czech Republic, June 25–29, 2012. Proceedings edited by F. Mika. Published by the CSMS (Brno, Czech Republic, 2012).

2014. 14th Seminar, incorporated into CPO-9, Brno, Czech Republic; see section 5.1.

# 6. SOME OTHER CONGRESSES

## 6.1 Introductory Notes

Internet has made it much simpler to identify proceedings issues and volumes than when I compiled the original document (Hawkes, 2003). Information about national meetings can be found on the society websites [see IFSM (ifsm.info), where links to the member countries of the European Microscopy Society (EMC), the Committee of Asia-Pacific Societies for Microscopy (CAPSM), and the Comité Interamericano de Sociedades de Microscopía, Interamerican Committee of Societies for Microscopy (CIASEM) can be found]. In this section, details are provided of the EMAG conferences, Scanning Microscopies, the two main series of Russian conferences, and conferences held by the Microscopy Society of Southern Africa. These have been selected somewhat arbitrarily because EMAG and the MSSA still produce full proceedings, while the Russian material can be elusive, although the contents lists of the *Bulletin of the*

*Russian Academy of Sciences (Physics)* now can be consulted on Springerlink (2007 onward), and some earlier issues of *Izvestiya* are accessible on www. gpi.ru/izvestiyaran-fiz. The series of workshops on LEEM and PEEM were not included in my earlier lists and therefore are included here, as are Frontiers of Aberration-Corrected Microscopy and the Polish International Conferences on Electron Microscopy of Solids.

## 6.2 Electron Microscopy and Analysis Group (EMAG), Institute of Physics

The proceedings of the biennial conferences of the Electron Microscopy and Analysis Group (EMAG) of the Institute of Physics are now published in the open-access journal *Journal of Physics: Conferences*. Printed copies are also supplied to participants in the meetings. Details of earlier meetings for which there were no formal proceedings are to be found in Hawkes (2003).

EMAG, 1971. Electron Microscopy and Analysis. Proceedings of the 25th Anniversary Meeting of the Electron Microscopy and Analysis Group of the Institute of Physics, Cambridge, UK, June 29–July 1, 1971 (Nixon, W. C., Ed.); London: Institute of Physics (1971); Conference Series *10*.

EMAG, 1973. Scanning Electron Microscopy: Systems and Applications, Newcastle-upon-Tyne, UK, July 3–5, 1973 (Nixon, W.C., Ed.); London: Institute of Physics (1973), Conference Series *18*.

EMAG, 1975. Developments in Electron Microscopy and Analysis. Proceedings of EMAG 75, Bristol, UK, September 8–11, 1975 (Venables, J. A., Ed.); London and New York: Academic Press (1976).

EMAG 1977. Developments in Electron Microscopy and Analysis. Proceedings of EMAG 77, Glasgow, September 12–14, 1977 (Misell, D. L., Ed.); Bristol, UK: Institute of Physics (1977); Conference Series *36*.

EMAG, 1979. Electron Microscopy and Analysis, 1979. Proceedings of EMAG 79, Brighton, UK, September 3–6, 1979 (Mulvey, T., Ed.); Bristol, UK: Institute of Physics (1980); Conference Series *52*.

EMAG, 1981. Electron Microscopy and Analysis, 1981. Proceedings of EMAG 81, Cambridge, UK, September 7–10, 1981 (Goringe, M. J., Ed.); Bristol, UK: Institute of Physics (1982); Conference Series *61*.

EMAG, 1983. Electron Microscopy and Analysis, 1983. Proceedings of EMAG 83, Guildford, UK, August 30–September 2, 1983 (Doig, P., Ed.); Bristol, UK: Institute of Physics (1984); Conference Series *68*.

EMAG, 1985. Electron Microscopy and Analysis, 1985. Proceedings of EMAG 85. Newcastle-upon-Tyne, UK, September 2–5, 1985 (Tatlock, G. J., Ed.); Bristol, UK: Institute of Physics (1986); Conference Series *78*.

EMAG, 1987. Electron Microscopy and Analysis, 1987. Proceedings of EMAG 87, Manchester, UK, September 8–9, 1987 (Brown, L. M., Ed.); Bristol, UK, and Philadelphia: Institute of Physics (1987); Conference Series *90*.

EMAG, 1989. EMAG-MICRO 89. Proceedings of the Institute of Physics Electron Microscopy and Analysis Group and Royal Microscopical Society Conference, London, September 13–15, 1989 (Goodhew, P. J., & Elder, H. Y., Eds.); Bristol, UK, and New York: Institute of Physics (1990); Conference Series *98*, 2 vols.

EMAG, 1991. Electron Microscopy and Analysis 1991. Proceedings of EMAG 91, Bristol, UK, September 10–13, 1991 (Humphreys, F. J., Ed.); Bristol, UK, Philadelphia, and New York: Institute of Physics (1991); Conference Series *119*.

EMAG, 1993. Electron Microscopy and Analysis 1993. Proceedings of EMAG 93, Liverpool, UK, September 15–17, 1993 (Craven, A. J., Ed.); Bristol, UK, Philadelphia, and New York: Institute of Physics (1994); Conference Series *138*.

EMAG 1995. Electron Microscopy and Analysis 1995. Proceedings of EMAG 95. Birmingham, UK, September 12–15, 1995 (Cherns, D., Ed.); Bristol, UK, Philadelphia, and New York: Institute of Physics (1995); Conference Series *147*.

EMAG 1997. Electron Microscopy and Analysis 1997. Proceedings of EMAG 97, Cavendish Laboratory, Cambridge, UK, September 2–5, 1997 (Rodenburg, J. M., Ed.; Bristol, UK, and Philadelphia: Institute of Physics (1997); Conference Series *153*.

EMAG 1999. Electron Microscopy and Analysis 1999. Proceedings of EMAG 99, University of Sheffield, 25–27 August 1999 (Kiely, C. J., Ed.); Bristol, UK, and Philadelphia: Institute of Physics (1999); Conference Series *161*.

EMAG 2001. Electron Microscopy and Analysis 2001. Proceedings of the Institute of Physics Electron Microscopy and Analysis Group Conference, University of Dundee, Dundee, Scotland, September 5–7, 2001 (Aindow, M., & Kiely, C. J., Eds.); Bristol, UK, and Philadelphia: Institute of Physics Publishing (2002); Conference Series *168*.

EMAG 2003. Electron Microscopy and Analysis 2003. Proceedings of the Institute of Physics Electron Microscopy and Analysis Group Conference, Examination Schools, University of Oxford, September 3–5, 2003 (McVitie, S., & McComb, D., Eds.); Bristol, UK, and Philadelphia: Institute of Physics Publishing (2004); Conference Series *179*.

EMAG–NANO 2005. University of Leeds, Leeds, UK, August 31–September 2, 2005 (Brown, P. D., Baker, R., & Hamilton, B., Eds.). *Journal of Physics: Conferences 26* (2006).

EMAG 2007. Caledonian University and University of Glasgow, 3–7 September 2007 (Baker, R. T., Möbus, G. & Brown, P. D., Eds.). *Journal of Physics: Conferences 126* (2008).

EMAG 2009. University of Sheffield, Sheffield, UK, September 8–11, 2009 (Baker, R. T., Ed.). *Journal of Physics: Conferences 241* (2010).

EMAG 2011. University of Birmingham, Birmingham, UK (Baker, R. T., Brown, P. D., & Li, Z., Eds.). *Journal of Physics: Conferences 371* (2012).

EMAG 2013. University of York, York, UK, September 3–6, 2013 (Nellist, P., Ed.). *Journal of Physics: Conferences 522* (2014).

EMAG 2015, Manchester, UK, June 29–July 2, 2015, joint with the Microscience Microscopy Conference (Royal Microsopical Society).

## 6.3 Southern Africa

The (Electron) Microscopy Society of Southern Africa continues to publish a *Proceedings* book in which the abstracts of the annual meetings are presented. (*MSSA-n* indicates that this is the *n*th meeting of the society.)

MSSA-43. Cape Town, South Africa, December 3–5, 2003. *MSSA Proceedings 33* (2003); Cowan, D.A., Marcus, K., Elliot, E., Kruger, H. & Cross, R. H. M. (Eds.).

MSSA-43. University of Pretoria, South Africa, December 1–3, 2004. *MSSA Proceedings 34* (2004); Cowan, D. A., Marcus, K., Theron, J., Kruger, H. & Cross, R. H. M. (Eds.).

MSSA-44. Pietermaritzburg, South Africa, December 5–7, 2005. *MSSA Proceedings 35* (2005); Aveling, T. A. S., Marcus, K., Theron, J., Wesley-Smith, J., & Sewell, B. T. (Eds.).

MSSA-45. Nelson Mandela Metropolitan University, Port Elizabeth, South Africa, November 29–December 1, 2006. *MSSA Proceedings 36* (2006); Aveling, T.A.S., Knutsen, R., Theron, J., Wesley-Smith, J. & Sewell, B.T., Eds.

Two-day Workshop only at Mintek in Randburg, South Africa, December 6–7, 2007, in view of the subsequent meetings in July 2008.

MSSA-46. Gaborone, Botswana, July 23–25, 2008. *MSSA Proceedings 38* (2008); Aveling, T. A. S., Knutsen, R., Theron, J., Wesley-Smith, J. & Sewell, B. T. (Eds.).

MSSA-47. University of Kwa-Zulu Natal, South Africa, December 8–11, 2009. *MSSA Proceedings 39* (2009); Aveling, T. A. S., Knutsen, R. & Sewell, B. T. (Eds.).

MSSA-48. Convention Centre of Forever Resorts in Warmbaths (Bela-Bela), University of Limpopo, South Africa, October 26–29, 2010. *MSSA Proceedings 40* (2010); Wesley-Smith, J., Knutsen, R. & Sewell, B. T. (Eds.).

MSSA-49. CSIR International Convention Centre, Pretoria, South Africa, December 6–9, 2011. *MSSA Proceedings 41* (2011); Baker, C., Knutsen, R. & Sewell, B. T. (Eds.).

MSSA-50. University of Cape Town, South Africa, December 4–7, 2012. *MSSA Proceedings 42* (2012); Baker, C., Neethling, J. H., & Sewell, B. T. (Eds.).

MSSA-51. Farm Inn, Pretoria, South Africa, December 3–6, 2013. *MSSA Proceedings 43* (2013); Loos, B., Westraadt, J. E., & Engelbrecht, J. A. A. (Eds.).

MSSA-52: Protea Hotel, Stellenbosch, 2–5 December 2014. *MSSA Proceedings 44* (2014); Loos, B., Westraadt, J.E. & Engelbrecht, J.E.E., (Eds.).

## 6.4 Russia

### 6.4.1 Russian Conferences on Electron Microscopy

Proceedings of the 18th Russian Conference on Electron Microscopy, Chernogolovka, Russia, June 5–8, 2000; *Izvestiya Rossiiskoi Akademii Nauk (Seriya Fizicheskaya)* or *Bulletin of the Russian Academy of Science: Physics, 65*(9) (2001).

Proceedings of the 19th Russian Conference on Electron Microscopy, Chernogolovka, Russia, May 27–31, 2002; *Izvestiya Rossiiskoi Akademii Nauk (Seriya Fizicheskaya)* or *Bulletin of the Russian Academy of Science: Physics, 67*(4) (2003).

Proceedings of the 20th Russian Conference on Electron Microscopy, Chernogolovka, Russia, June 2004; *Izvestiya Rossiiskoi Akademii Nauk (Seriya Fizicheskaya)* or *Bulletin of the Russian Academy of Science: Physics, 69*(4) (2005).

Proceedings of the 21st Russian Conference on Electron Microscopy, Chernogolovka, Russia, June 5–10, 2006; *Izvestiya Rossiiskoi Akademii Nauk (Seriya Fizicheskaya)* or *Bulletin of the Russian Academy of Science: Physics, 71*(10) (2007).

Proceedings of the 22nd Russian Conference on Electron Microscopy, Chernogolovka, Russia, 2008; *Izvestiya Rossiiskoi Akademii Nauk (Seriya Fizicheskaya)* or *Bulletin of the Russian Academy of Science: Physics,* *73*(4) (2009); also *Poverkhnost'* No. 10 (2009); *Journal of Surface Investigation X-Ray Synchrotron Neutron Techniques,* *3*(5) (2009).

Proceedings of the 23rd Russian Conference on Electron Microscopy, Chernogolovka, Russia, 2010; *Izvestiya Rossiiskoi Akademii Nauk (Seriya Fizicheskaya)* or *Bulletin of the Russian Academy of Science: Physics,* *75*(9) (2011); also *Poverkhnost'* (No. 10) (2011).

Proceedings of the 24th Russian Conference on Electron Microscopy, Chernogolovka, Russia, 2012; *Izvestiya Rossiiskoi Akademii Nauk (Seriya Fizicheskaya)* or *Bulletin of the Russian Academy of Science: Physics,* *77*(8) (2013).

Proceedings of the 25th Russian Conference on Electron Microscopy, Chernogolovka, Russia, 2014; *Izvestiya Rossiiskoi Akademii Nauk (Seriya Fizicheskaya)* or *Bulletin of the Russian Academy of Science: Physics,* *79* (2015).

### 6.4.2 Symposia on Scanning Electron Microscopy and Analytical Methods in the Study of Solids[2]

Proceedings of the Eleventh National Symposium on Scanning Electron Microscopy and Analytical Methods in the Study of Solids, SEM–99, Chernogolovka, Russia, May 1999. *Izvestiya Rossiiskoi Akademii Nauk (Seriya Fizicheskaya)* or *Bulletin of the Russian Academy of Science: Physics,* *64*(8) (2000); *Poverkhnost'* No. 12 (2000).

Proceedings of the Twelfth National Symposium on Scanning Electron Microscopy and Analytical Methods in the Study of Solids, SEM–2001, Chernogolovka, Russia, June 4–6, 2001. *Izvestiya Rossiiskoi Akademii Nauk (Seriya Fizicheskaya)* or *Bulletin of the Russian Academy of Science: Physics,* *66*(9) (2002).

Proceedings of the Thirteenth National Symposium on Scanning Electron Microscopy and Analytical Methods in the Study of Solids, SEM–2003, Chernogolovka, Russia, May 2003. *Izvestiya Rossiiskoi Akademii Nauk (Seriya Fizicheskaya)* or *Bulletin of the Russian Academy of Science: Physics,* *68*(9) (2004); *Poverkhnost'* No. 3 (2004).

Proceedings of the Fourteenth National Symposium on Scanning Electron Microscopy and Analytical Methods in the Study of Solids, SEM–2005, Chernogolovka, Russia, May 30–June 3, 2005. Not found in *Izvestiya*

---

[2] For the contents of *Poverkhnost*, see www.issp.ac.ru/journal/surface.

*Rossiiskoi Akademii Nauk (Seriya Fizicheskaya)* or *Bulletin of the Russian Academy of Science: Physics, 70* (2006); *Poverkhnost'* No. 9 (2006).

Proceedings of the Fifteenth National Symposium on Scanning Electron Microscopy and Analytical Methods in the Study of Solids, SEM–2007, Chernogolovka, Russia, 2007. *Izvestiya Rossiiskoi Akademii Nauk (Seriya Fizicheskaya)* or *Bulletin of the Russian Academy of Science: Physics, 72*(11) (2008); *Poverkhnost* (No. 9) (2008), *Journal of Surface Investigation X-Ray Synchrotron Neutron Techniques, 2*(5) (2008).

Proceedings of the Sixteenth National Symposium on Scanning Electron Microscopy and Analytical Methods in the Study of Solids, SEM–2009, Chernogolovka, Russia, 2009. *Izvestiya Rossiiskoi Akademii Nauk (Seriya Fizicheskaya)* or *Bulletin of the Russian Academy of Science: Physics, 74*((2010) No. 7; *Poverkhnost*(9) (2010), Not found in *Journal of Surface Investigation X-Ray Synchrotron Neutron Techniques.*

Proceedings of the Seventeenth National Symposium on Scanning Electron Microscopy and Analytical Methods in the Study of Solids, SEM–2011, Chernogolovka, Russia, 2011. *Izvestiya Rossiiskoi Akademii Nauk (Seriya Fizicheskaya)* or *Bulletin of the Russian Academy of Science: Physics, 76* (No. 9) (2012); not found in *Poverkhnost* (2012) or in *Journal of Surface Investigation X-Ray Synchrotron Neutron Techniques.*

Proceedings of the Eighteenth National Symposium on Scanning Electron Microscopy and Analytical Methods in the Study of Solids, SEM–2013, Chernogolovka, Russia, 2013. *Izvestiya Rossiiskoi Akademii Nauk (Seriya Fizicheskaya)* or *Bulletin of the Russian Academy of Science: Physics, 78*(9) (2014).

## 6.5 International Congresses on X-Ray Optics and Microscopy (ICXOM)

ICXOM XVI, Proceedings of the 16th International Congress on X-Ray Optics and Microanalysis, Vienna University of Technology, July 2–6, 2001. Papers submitted either to *Journal of Analytical Atomic Spectroscopy* or to *Spectrochimica Acta B, 58*(4) (2002) (ICXOM issue); Mantler, M., Wobrauschek, P., Friedbacher, G. & Schreiner, M. (Eds.).

ICXOM XVII, Proceedings of the 17th International Congress on X-Ray Optics and Microanalysis, Chamonix, France, September 22–26, 2003. *Spectrochimica Acta B, 59*(10–11) (2004) (ICXOM issue); Simionovici, A. (Ed.).

ICXOM-18, Proceedings of the 18th International Congress on X-Ray Optics and Microanalysis, Frascati, Italy, September 25–30, 2005. *Spectrochimica Acta B, 62*(6–7) (2006) (ICXOM issue); Dabagov, S. (Ed.).

ICXOM-19, Proceedings of the 19th International Congress on X-Ray Optics and Microanalysis, Kyoto, Japan, September 16–21, 2007. *Spectrochimica Acta B, 64*(8) (2009) (ICXOM issue); Kawai, J., & Janssens, K. (Eds.)

ICXOM-20, Proceedings of the 20th International Congress on X-Ray Optics and Microanalysis, Forschungszentrum Karlsruhe, Leopoldshafen, Germany, September 15–18, 2009. AIP Conference Proceedings *1221* (2010); Denecke, M. A., & Walker, C. T. (Eds.). New York and Heidelberg, Germany: Springer.

ICXOM-21, Proceedings of the 21st International Congress on X-Ray Optics and Microanalysis Campinas, Brazil, September 5–8, 2011. AIP Conference Proceedings *1437* (2012); Perez, C.A. & Malachias de Souza, A. (Eds.). Distributed by Springer (New York and Heidelberg).

ICXOM-22, Proceedings of the 22nd International Congress on X-Ray Optics and Microanalysis, Hamburg, Germany, September 2–6, 2013. *Journal of Physics: Conferences. 499* (2014) (Falkenberg, G., & Schroer, C., Eds.).

ICXOM-23, Brookhaven National Laboratory, Upton, NY, September 14–18, 2015.

## 6.6 Scanning

The proceedings of these annual meetings, formerly published as short abstracts in *Scanning*, are now published as volumes of the SPIE Conference Proceedings series.

Scanning 2003. Doubletree Hotel, San Diego Mission Valley, CA, May 3–5, 2003. *Scanning, 25*(2) (2003); Becker, R. P. (Ed.).

Scanning 2004. Washington, DC, April 27–29, 2004. *Scanning, 26*(2) (2004); Becker, R. P. (Ed.).

Scanning 2005. Monterey, CA, April 5–7, 2005. *Scanning, 27*(2) (2005); Becker, R. P. (Ed.).

Scanning 2006. Washington, DC, April 25–27, 2006. *Scanning, 28*(2) (2006); Becker, R. P. (Ed.).

Scanning 2007. Monterey, CA, April 10–12, 2007. *Scanning, 29*(2) (2007); Becker, R. P. (Ed.).

Scanning 2008. Gaithersburg, MD, April 15–17, 2008. *Scanning, 30*, 240–282 (2008); Becker, R. P. (Ed.).

Scanning Microscopy 2009. Monterey, CA, May 4–7, 2009. *Proceedings SPIE 7378* (2009); Postek, M. T., Newbury, D. E., Platek, S. F., & Joy, D. C. (Eds.).

Scanning Microscopy 2010. Monterey, CA, May 17–19, 2010. *Proceedings SPIE 7729* (2010); Postek, M. T., Newbury, D. E., Platek, S. F., & Joy, D. C. (Eds.).

Scanning Microscopies 2011. Orlando, FL, April 26–28, 2011. *Proceedings SPIE 8036* (2011); Postek, M. T., Newbury, D. E., Platek, S. F., Joy, D. C., & Maugel, T.K. (Eds.).

Scanning Microscopies 2012. Advanced Microscopy Technologies for Defense, Homeland Security, Forensic, Life, Environmental, and Industrial Sciences. Baltimore, April 24–26, 2012. *Proceedings SPIE 8378* (2012); Postek, M. T., Newbury, D. E., Platek, S. F., Joy, D. C., & Maugel, T. K. (Eds.).

Scanning Microscopies 2013. Advanced Microscopy Technologies for Defense, Homeland Security, Forensic, Life, Environmental, and Industrial Sciences. Baltimore, April 30–May 1, 2013. *Proceedings SPIE 8729*. Postek, M. T., Newbury, D. E., Platek, S. F., & Maugel, T. K. (Eds.).

Scanning Microscopies 2014. Monterey, CA September 16–18, 2014. *Proceedings SPIE 9236* (2014). Postek, M. T., Newbury, D. E., Platek, S. F., & Maugel, T.K. (Eds.).

## 6.7 Low-Energy Electron Microscopy and Photoemission Electron Microscopy (LEEM, PEEM)

Tempe, AZ, April 7–9, 1998. *Surface Review Letters,* 5(6), 1129–1136 (1998).

Paris, September 26–28, 2000. *Surface Science,* 480(3), 97–218 (2001).

Albuquerque, NM, May 14–17, 2002. *Journal of Vacuum Science & Technology B,* 20(6), 2472–2550 (2002).

University of Twente, Enschede, the Netherlands, May 10–13, 2004. *Journal of Physics: Condensed Matter,* 17(16), S1305–S1426 (2005).

Himeji, Japan, October 15–19, 2006. *Surface Science,* 601(20), 4163–4773 (2007).

Trieste, Italy, September 7–11, 2008. *Journal of Physics: Condensed Matter,* 21(31), 310301–315005 (2009).

New York, August 8–13, 2010. *IBM Journal of Research and Development,* 55(4) (2011).

Hong Kong, November 11–15, 2012. *Ultramicroscopy 130,* 1–114 (2013).

Berlin, September 14–18, 2014. *Ultramicroscopy* (2015).

## 6.8 Frontiers of Aberration-Corrected Electron Microscopy

PICO-2012. Jülich, Germany, February 29–March 2, 2012.

PICO-2013. Kasteel Vaalsbroeck, Vaals, Netherlands, October 9–12, 2013. *Ultramicroscopy 134* (2013).

PICO-2015. Kasteel Vaalsbroeck, Vaals, Netherlands, April 19–23, 2015. *Ultramicroscopy. 151* (2015).

## 6.9 Polish International Conferences on Electron Microscopy of Solids

The proceedings of the first 10 conferences were prepared by the conference organizers. Those of subsequent meetings were published in various journals, as indicated here:

Gliwice, 1963

Warszawa, 1971

Kraków, 1973

Gliwice-Wisla, 1975

Warszawa-Jadwisin, 1978

Kraków-Krynica, 1981

Kraków-Krynica, 1989

Wroclaw-Szklarska, 1993

Kraków-Zakopane, 1996

Warszawa-Serock, 1999

Kraków-Krynica, 2002. *Materials Chemistry and Physics, 81*(2–3) (2003).

Warszawa-Kazimierz Dolny, 2005. *Journal of Microscopy, 223*(3) (2006); and *224*(1) (2006).

Kraków-Zakopane, 2008. *Journal of Microscopy 237*(3) (2010).

Wisla, 2011. *Electron Microscopy XIV. Solid-State Phenomena* (Trans Tech Publications, 2012).

Cracow, September 15–18, 2014. Many of the papers presented will be printed in *International Journal of Materials Research, Solid-State Phenomena,* or *Folia Histochemica & Cytobiologica.*

## Reference

Hawkes, P. W. (2003). Electron optics and electron microscopy: Conference proceedings and abstracts as source material. *Advances in Imaging and Electron Physics, 127,* 207–379.

CHAPTER FOUR

# Scanning Thermal Microscopy (SThM): How to Map Temperature and Thermal Properties at the Nanoscale

## Grzegorz Wielgoszewski[a,b,*], Teodor Gotszalk[a]

[a]Wrocław University of Technology, Faculty of Microsystem Electronics and Photonics, Nanometrology Group, ul. Z. Janiszewskiego 11/17, PL-50372 Wrocław, Poland
[b]University College Dublin, School of Physics, Science Centre North, Belfield, Dublin 4, Ireland
*Corresponding author: e-mail address: Grzegorz.Wielgoszewski@pwr.edu.pl

## Contents

*Advances in Imaging and Electron Physics*, Volume 190
ISSN 1076-5670
http://dx.doi.org/10.1016/bs.aiep.2015.03.011

# 1. INTRODUCTION

Various methods of investigation of microstructures and nanostructures, which are now known as *scanning probe microscopy (SPM) methods,* began to be developed over 30 years ago. Nowadays, a vast number of variations of the original scanning tunneling microscopy (STM) serve to collect information not only on the nanoscale shape of the surface but also on its physical and chemical properties (see also the section entitled "SPM," later in this chapter). Generally, qualitative imaging has been enabled, revealing specific contrasts across the investigated areas. In most cases, such information is satisfactory; however, the quota of applications that require quantitative data has recently increased.

The main goal of this chapter is to describe scanning thermal microscopy (SThM), which has been designed for the investigation of thermal properties at the nanoscale. Additionally, a brief introduction to various SPM methods is provided to show how SThM is situated among many specialized SPM-based techniques.

# 2. SPM

In the 1970s, Russell Young, John Ward, and Fredric Scire of the U.S. National Bureau of Standards (NBS), which is now known as the National Institute of Standards and Technology (NIST), demonstrated the Topografiner, an innovative microprofilometer (Young, Ward, & Scire, 1972; Villarrubia, Scire, Teague, & Gadzuk, 2001). The device allowed imaging of the topography of an electrically conductive surface by the use of electron field emission, incorporating concepts published in Young (1966). Using ideas similar to those behind the Topografiner, Gerd Binnig, Heinrich Rohrer, Christoph Gerber, and Eduard Weibel of IBM Zurich Research Laboratory at Rüschlikon, Switzerland, recorded in 1981 the first-ever image of a surface in which single atoms were distinguishable

(Binnig, Rohrer, Gerber, & Weibel, 1982; Binnig & Rohrer, 1987). Five years later, "for their design of the scanning tunnelling microscope" (Nobelprize.org, 1986), Binnig and Rohrer were awarded the Nobel Prize in Physics, which they shared with Ernst Ruska, who presented the transmission electron microscope in 1933.

The STM built by Binnig et al. (1982) operated based on the ability of an electron to "tunnel" through a potential barrier that separates two regions, a phenomenon that is derived from quantum mechanics. Crucial for the usefulness of this effect was the strong dependence of the tunneling current on the barrier width, which in the case of STM is constituted by the vacuum gap (or an air gap) between the tip and the sample. The exponential relationship facilitates atomic-scale measurements; however, there is also a limitation: a correct surface image is recorded only if the surface work function is constant.

In December 1985, a year before he received the Nobel Prize for the STM, Binnig, together with Calvin Quate and Christoph Gerber, submitted a paper on an atomic force microscope (Binnig, Quate, & Gerber, 1986). In the experiments described therein, the STM was used to measure the displacement of a spring microcantilever. Having a sharp tip on its end and being scanned across a surface, such a microcantilever enabled the recording of an image of a nonconductive surface in air, providing resolution of 3 nm along the $x$- and $y$-axes and below 0.1 nm along the $z$-axis.

As opposed to an STM result, a basic topographical atomic force microscopy (AFM) result is independent of electrical properties of the surface. The obtained image is based on the forces present between the tip and the sample, which include such interactions as general repulsive intermolecular forces, capillary forces, or van der Waals forces, as well as magnetic, electrostatic, or chemical ones. Their values range from $10^{-12}$ N to $10^{-7}$ N (Gotszalk, 2004; Haugstad, 2012), while the imaging limitations are mostly connected with the availability of suitable microprobes and precision electronic devices. Usually, the measurement microprobe, which acts as a force transducer, is a spring microcantilever with a sharp tip on its free end. The force measurement is realized by approaching the microprobe to the surface and detecting the cantilever deflection.

The "eye" of the scanning probe microscope is its sharp-tipped microprobe. These are the properties of the probe tip that determine the measurement possibilities, including the fact that it is mostly the tip shape that determines the spatial resolution. To give an idea how the microprobe may look, a typically shaped microcantilever with a tip is shown in

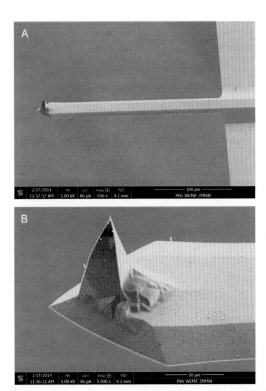

**Figure 1** A typically shaped SPM microprobe, the Pointprobe CDT-FMR by NanoWorld AG, made of silicon and coated with diamond thin film, designed for conductive AFM: (A) Microcantilever with tip, (B) Close-up of the tip (Wielgoszewski et al., 2008; NanoWorld AG, 2015). Microcantilever dimensions: 225 $\mu$m × 28 $\mu$m × 3 $\mu$m; tip height: about 15 $\mu$m, resonant frequency: 75 kHz; spring constant: 2.8 N/m. Scanning electron microscope (SEM) images courtesy of Karolina Orłowska (Wrocław University of Technology).

Figure 1. Such microprobes are a result of silicon-etching processes, while the surface of the one shown in the micrograph is also covered with a diamond layer to facilitate current measurements. As can be seen, the SPM microprobe is usually a rectangular cantilever, typically with a length of 100–500 $\mu$m, width of 20–50 $\mu$m, and a thickness of 1–3 $\mu$m. The pyramidal tip is usually of height of 10–15 $\mu$m.

Another feature common to almost all SPM systems is the ability to scan across the surface, which in this case means relative motion of the probe and the investigated sample. Eventually, an image is recorded, every pixel of which represents the desired tip–sample interactions in the corresponding point. In most cases, such scanning is realized by using a set of piezoelectric

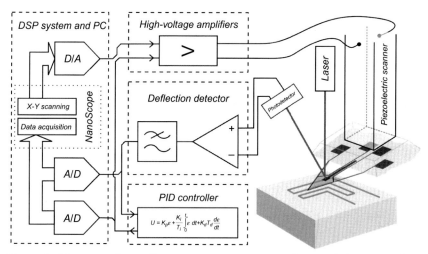

**Figure 2** Schematic of an AFM measurement system with optical detection of the deflection of the cantilever, in which it is the microprobe that is mounted on the piezoelectric scanner (see also Wielgoszewski et al., 2011b). (See the color plate.)

devices that provide movement in all three axes. Items to be scanned are placed either under the sample or above the microprobe (e.g., as shown in Figure 2), depending on the specific design of the SPM system.

The developments that followed those made by Binnig and his collaborators can be generally divided into three groups. Some of them were intended to increase the imaging resolution, which resulted in showing even sub-angstrom details (Hembacher, Giessibl, & Mannhart, 2004). Others were meant to increase the general measurement capabilities, which lead to simultaneous imaging of surface topography and various properties. There are also methods that use SPM-based solutions but are not used for scanning the surface, being a sophisticated gas or liquid sensing system. In the following section, some of these methods are briefly introduced.

## 2.1 Basic SPM Techniques

SPM methods can be classified in various ways. The most general one is to consider the type of phenomenon that is involved in recording any type of image, usually beginning with the surface shape (i.e., the topography). The four basic phenomena that enable all known SPM techniques are the following:

- Tunneling current
- Molecular forces

- Heat flow
- Electromagnetic waves

The existence of tunneling current makes STM possible. The molecular forces give the opportunity to develop a variety of techniques based on AFM. Without the heat flow, scanning near-field thermal microscopy (SNThM) would not be available, while the operation of scanning near-field optical microscopy (SNOM) is based on electromagnetic waves. These are only examples, and it is worth underlining that one advanced SPM technique may use a few different phenomena; e.g., tunneling shear force microscopy (T-ShFM) uses a shear force sensor with a conductive tip, thus combining the first two phenomena (Woszczyna et al., 2010a).

The term *molecular force* has probably the broadest meaning of the four abovementioned phenomena. Indeed, the AFM has evolved into countless methods, as it is limited only by the creativity of AFM users, who keep finding new applications of existing modules and the inventiveness of designers of new microprobes and measurement systems. However, keeping the force as the essential factor of the AFM operation, four basic modes can be distinguished:

- Contact (static) AFM, in which the repulsive interactions between a deflected cantilever and the surface are used
- Noncontact atomic force microscopy (NC-AFM), in which there are used only the attractive forces between an oscillating cantilever and the surface
- Intermittent contact AFM (also known as *tapping-mode AFM*), in which the tip periodically hits the surface and the imaging is based on the repulsive forces
- Shear force microscopy (ShFM), in which the considered forces act perpendicularly to the direction of cantilever oscillations

All these modes in the basic setup enable imaging the surface topography [i.e., the true three-dimensional (3D) shape of the surface]. If an advanced setup is used, a simultaneous image of certain additional properties can be obtained (see the section entitled "Advanced SPM Techniques," later in this chapter).

Another feature that determines the capabilities of an AFM system is the way the output signal of the force transducer is detected. The majority of AFM setups include a microcantilever with a sharp tip, the deflection of which may be transformed into an electrical signal in the following ways (Michels & Rangelow, 2014):

- Using tunneling current; which was the original method (Binnig et al., 1986) but is hardly used today

- Using a position-sensitive photodiode, also known as a position-sensitive detector (PSD), which detects the position of a laser beam reflected from the cantilever (Meyer & Amer, 1988, 1990)
- Using an interferometer to measure the deflection directly (Rugar, Mamin, & Güthner, 1989)
- Using piezoresistors embedded in the cantilever during its manufacture (Linnemann, Gotszalk, Rangelow, Dumania, & Oesterschulze, 1996; Gotszalk, Grabiec, & Rangelow, 2000; Woszczyna et al., 2010b)

All these techniques enable to express the interaction force between the tip and the surface as an electrical signal, which is further amplified, processed, and recorded. The PSD-based method is currently the most popular, as in a relatively simple manner (by using a low-powered laser, mirrors, and a four-sectional PSD), it provides both deflection and torsion signals, which can be later used in topography and friction imaging, among other purposes.

Most AFM systems incorporate a proportional–integral–derivative (PID) controller. Such a device enables operation in feedback mode by moving the piezoelectric scanner to the $z$-axis, so that the tip–sample force is kept constant. As a result, the inverted PID output signal is interpreted as the surface shape. In Figure 2, a basic AFM setup is shown that schematically presents all the devices that are necessary for basic AFM operation.

Taking into account the three abovementioned lists of factors, it is now easier to imagine how many variants of AFMs are possible. It should be noted that only the basic topography imaging has been described so far. Generally, to any combination of the ideas given previously, a dedicated extension module can be added, enabling the measurement of more surface (or even sub-surface) properties. A tiny fraction of all possible AFM variants, that had been developed over 30 years in laboratories worldwide, is presented in the next section.

## 2.2 Advanced SPM Techniques

So far as the variety of SPM techniques is concerned, there is another important feature of most SPM systems: their modular design. A user who owns a basic AFM system can improve it fairly easily by adding new dedicated or universal modules. Sometimes even buying a new type of microprobes may allow operation in an additional mode. However, in most cases, dedicated electronic devices are also needed.

Most commercially available AFM systems allow the user to operate in the following advanced modes:

- Advanced mechanical AFM modes, in which the mechanical properties of the probe are used to gather more information than just the surface shape:

- Friction force microscopy (FFM) or lateral force microscopy (LFM), in which recording the cantilever twist enables imaging the friction forces between the tip and the surface
- Force modulation microscopy (FMM), in which the load force of the tip on the surface is modulated, so that information (such as elastic properties of the sample) can be collected
- Electrical AFM modes (Avila & Bhushan, 2010), in which the tip is made of an electrically conductive material and its potential can be controlled:
  - Electrostatic force microscopy (EFM), enabling the recording of the surface charge or potential variations based on changes in the forces acting on the tip (Terris, Stern, Rugar, & Mamin, 1989; Girard, 2001);
  - Kelvin probe force microscopy (KPFM), enabling the tracking of the surface potential, maintaining a constant distance between the tip and the surface (Nonnenmacher, O'Boyle, & Wickramasinghe, 1991; Bhushan & Goldade, 2000; Moczała, Sosa, Topol, & Gotszalk, 2014)
  - Conductive(−probe) atomic force microscopy (C-AFM), in which the current flowing through the tip is recorded while the surface is scanned in static (contact) mode (Murrell et al., 1993; Wielgoszewski, Gotszalk, Woszczyna, Zawierucha, & Zschech, 2008; Gajewski et al., 2015)
  - Piezoresponse force microscopy (PFM), in which the tip acts as an electrode to excite deformation of a piezoelectric sample (Güthner & Dransfeld, 1992; Huey et al., 2004)
  - Scanning capacitance microscopy (SCM), which enables the recording of variations in tip–surface capacitance (Abraham, Williams, Slinkman, & Wickramasinghe, 1991; Lányi, 2008)
  - Scanning spreading resistance microscopy (SSRM), similar to C-AFM but usually with a higher current range, enabling investigations of doping levels in semiconductors (De Wolf, Snauwaert, Clarysse, Vandervorst, & Hellemans, 1995), among other uses
- Magnetic force microscopy (MFM), which uses a microprobe with a tip made of magnetic material (Sáenz et al., 1987).
- SThM, which adds the capability of imaging thermal properties of the sample

This list cannot be considered complete: there are a large variety of specialized scanning-based AFM modes, including those involving more than one cantilever being used at once (Sulzbach & Rangelow, 2010; Ivanova et al., 2008). It should also be mentioned here that many AFM-related techniques do not require scanning. These are used for local investigations of adhesion or elastic properties (in the case of tip-based measurements), but also for experiments carried out in liquid environments, such as when a chemically functionalized tipless cantilever can be used for the ultraprecise detection of specific substances (Ndieyira et al., 2008; Braun et al., 2009; Nieradka, Gotszalk, & Schroeder, 2012).

This brief introduction to scanning probe microscopy is followed by a much more detailed description of one of the above mentioned techniques: SThM.

## 3. SThM—THE DEVELOPMENT OF THE TECHNIQUE

### 3.1 The Origins of SThM

The origins of SThM can be found in the late 1980s, when Clayton C. Williams and H. Kumar Wickramasinghe of IBM T. J. Watson Research Center in Yorktown Heights, NY, published papers on the scanning thermal profiler (SThP) and high-resolution thermal microscopy (Williams & Wickramasinghe, 1986a, b). The operation principle of SThP was very similar to the STM system: the difference was that constant heat flow was used to maintain a constant tip–sample distance, rather than the tunneling current. The probe was made of a sharp tip with a nanothermocouple at its apex, forming the nanothermometer. In such a setup, a forced temperature difference and tip–sample heat exchange were used to image the surface topography. The authors stated that the thermal conductivity of any solid is much higher than that of air, and so the recorded topography would be independent of the material properties of the sample. This capability was quite important: at that time, the AFM was not well known yet. An advantage over both STM and AFM was also stated: a reasonable imaging resolution of a few hundred nanometers was achieved for the tip fly height of 0.1 $\mu$m above the sample, unlike in the two other techniques, where the distance needed to be much smaller (Williams & Wickramasinghe, 1986b).

Also in 1986, Williams and Wickramasinghe proposed the use of SThP to record the surface temperature (Williams & Wickramasinghe, 1986a). If the heat flow scanning feedback was turned off and tip temperature kept constant, a temperature map should be acquired, again based on the tip–sample heat flow rate. A few years later, localized temperature measurement was used for investigations of nanoscale light absorption (Weaver, Walpita, & Wickramasinghe, 1989). Another nanothermocouple-based system was utilized as a scanning chemical potential microscope, and the atomic structures of molybdenum disulfide and liquid crystals on graphite surface were investigated using that instrument. In the latter case, the thermoelectric signal enforced by temperature difference was used (Williams & Wickramasinghe, 1990).

SThM systems, in which simultaneous recording of topography and thermal images, have appeared in the 1990s. In 1993 Majumdar, Carrejo, and Lai prepared an AFM-probe equivalent, made of a chromel–alumel pair of wires (i.e., $Ni_{0.9}Cr_{0.1} - Ni_{0.95}AlMnSi$), with their junction forming a tip. In this way, an SThM microprobe with a K-type thermocouple was constructed, which was later used to record surface temperature maps independent of the surface topography recorded simultaneously (Majumdar, Carrejo, & Lai, 1993). This system, therefore, could be called the first scanning thermal microscope as it is understood nowadays.

Further development of SThM techniques was mostly related to designing new kinds of thermal probes. These are more thoroughly described in the section entitled "Thermal Probes for Use in SThM," later in this chapter. For now, it is sufficient to state that Dinwiddie, Pylkki, and West (1994) introduced the thermoresistive Wollaston microprobe. Its simple design, which enabled fairly repeatable manufacturing, led to commercialization of the SThM techniques, which was realized by TopoMetrix and Thermomicroscopes Inc.

## 3.2 SThM Since 1995

The significance of the Wollastone probe introduction can be seen in Figure 3, which shows the number of SThM-related papers between 1986 and 2012, based on the Thomson Reuters Web of Science database (Thomson Reuters, 2012). It can be clearly seen from this that the number of papers on SThM has notably increased since 1994. This rise is connected to the better accessibility of SThM, which started with its commercialization, which is also confirmed by the fact that in the most of these papers, a Wollaston probe was utilized. Still, however, thermocouple-based probes have been extensively developed in the laboratory. For example, very advanced thermocouple probes were designed by Majumdar's team from 1995–2001; it is this team that holds the current passive-mode SThM resolution record, having shown a temperature image of a multi-walled nanotube with sectional FWHM of 50 nm (Shi, Kwon, Majumdar, & Miner, 2001).

The many papers on applications of SThM published since 1995 have been accompanied by works on new probe types (Luo, Shi, Varesi, &

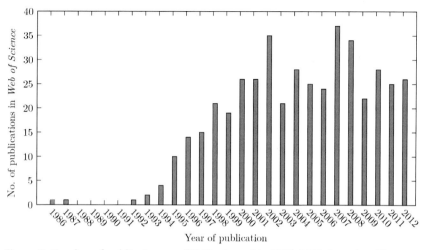

**Figure 3** Number of publications on SThM released in 1986–2012, based on Thomson Reuters Web of Science (Wielgoszewski, 2014). (See the color plate.)

Majumdar, 1997; Mills, Zhou, Midha, Donaldson, & Weaver, 1998; Edinger, Gotszalk, & Rangelow, 2001; Janus et al., 2010; Zhang, Dobson, & Weaver, 2011; Tovee, Pumarol, Zeze, Kjoller, & Kolosov, 2012; Janus et al., 2014; Hofer et al., 2015), calibration methods (Lefèvre, Saulnier, Fuentes, & Volz, 2004; Dobson, Mills, & Weaver, 2005; Wielgoszewski, Babij, Szeloch, & Gotszalk, 2014), and attempts to improve the SThM technique itself (Oesterschulze & Stopka, 1996; Pollock & Hammiche, 2001; Dobson, Weaver, & Mills, 2007; Kim et al., 2008; Wielgoszewski et al., 2011b; Juszczyk, Wojtol, & Bodzenta, 2013). To date, the greatest impact on the development of AFM-based thermal analysis comes from the following elements:

- Development of microthermal analysis ($\mu$TA) (Pollock & Hammiche, 2001) and the closely related development of doped-silicon probes (discussed in the section entitled "Thermoresistive SThM Probes," later in this chapter), which enabled localized calometric measurements. Nano-thermal analysis (nano-TA) systems, that utilize more advanced probes and provide better spatial resolution, have been commercially available since 2007.
- Introduction of the KNT-SThM thermoresistive nanoprobe (Dobson et al., 2007), which allowed much better imaging parameters than the Wollastone probe, being not only an improvement but also an incentive for those working on even more advanced designs. This has been commercially available since 2008.

- Development of a micromechanical calibration stage (Dobson et al., 2005), which so far offers the most accurate calibration of an SThM probe.

There also have been significant contributions to the analysis of tip–sample physical contact (e.g., Thiery, Toullier, Teyssieux, & Briand, 2008; Kim & King, 2009; Fletcher, Lee, & King, 2012; Gotsmann & Lantz, 2013; Tovee & Kolosov, 2013), including a paper introducing a novel technique meant to help compensating for contact effects during data postprocessing called *tip thermal mapping (TThM)* (Jóźwiak et al., 2013).

## 4. PRINCIPLES OF SThM OPERATION

SThM provides a means of simultaneous recording temperature or thermal conductivity maps along with surface topography. The information is gathered point by point, as the probe is being moved relative to the sample. Last but not least, the obtained signal is processed and forwarded to the measurement and control system. Having taken these into consideration, four top-level blocks may be distinguished (see Figure 4):

- A probe that includes a thermal sensor
- A device to process the temperature signal (generally referred to as an *SThM module*)
- A sample movement and topography measurement system
- An acquisition and control system

Depending on specific system guidelines, these blocks may be realized in various ways. Possible types of SThM probes are described in the section entitled "Thermal Probes for Use in SThM," later in this chapter. The temperature measurement devices, design of which may depend on probe used, are introduced in the section entitled "Measurement Devices for SThM,"

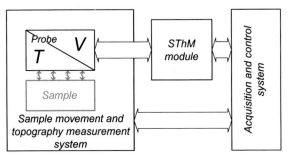

**Figure 4** Basic blocks comprising an SThM measurement system; the SThM probe is shown as a temperature-to-voltage converter.

**Figure 5** A simple differentiation between SThM operating modes: passive (P-SThM) and active (A-SThM). (See the color plate.)

later in this chapter. The other two blocks are a regular atomic force microscope, which formed a basis for today's scanning thermal microscope.

The main capability of an SThM system is the highly localized measurement of thermal phenomena in the sample. It starts with an atomic force microscope, which should generally enable imaging the surface shape with spatial resolution reaching 0.1 nm. If the standard AFM probe is replaced by a nanothermometer, two additional advanced measurement modes become available:

- Passive-mode scanning thermal microscopy (P-SThM), in which the probe acts as a thermometer
- Active-mode scanning thermal microscopy (A-SThM), in case the probe is both a thermometer and a heat source

Generally, in P-SThM, the heat flow direction depends on the tip-sample temperature difference, while in A-SThM, it does not change. As the P-SThM mode is usually used to localize hot spots, the heat flow in P-SThM can be characterized as sample $\Rightarrow$ probe; meanwhile, in A-SThM, it would be the opposite, probe $\Rightarrow$ sample (see Figure 5). Therefore, P-SThM would be used to obtain the temperature maps, while A-SThM would allow investigations of heat flow and, e.g., the thermal conductivity.

## 4.1 P-SThM

The word *passive* in the spellout of P-SThM refers to the fact that in this mode, the scanning process should not change the state of the sample. Rather, the measurement should result in recording the temperature map as it is generated by phenomena in the sample itself. Therefore, only the probe response to temperature changes is recorded. Also, it should be noted that such a mode of operation does not require any specific thermal probes: any nanothermometer would be enough.

In most cases, P-SThM is used for visualizing the in situ temperature maps in microelectronic and nanoelectronic systems. Its usefulness comes

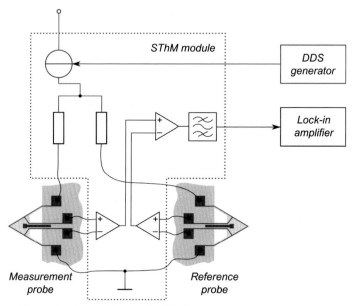

**Figure 6** Schematic of a passive-mode SThM measurement setup.

from the fact that the resolution offers a precise localization of "hot spots," the areas in which the current flow causes excessive heat generation. This may help to identify narrowing of the lines because of technological errors or electromigration, as well as provide diagnostics of on–chip resistive elements.

To increase the sensitivity, the SThM probe is usually placed in a measurement bridge. In Figure 6, a two–probe setup with a modified Wheatstone bridge topology is shown. One of the probes is used for scanning across the surface, while the other provides compensation for ambient temperature changes. More detailed description of various bridge setups is provided in the section entitled "Measurement Devices for SThM," later in this chapter.

## 4.2 A-SThM

Contrary to P-SThM, the word *active* in the name of A-SThM is used to indicate that there is an excitation coming from the measurement system, which allows for observing thermal properties of the sample regardless of heat generation occurring in it. In most A-SThM systems, the tip is heated by a high-density current flowing through it. This additional feature does not affect the P-SThM-like temperature measurement; therefore, two variants of A-SThM are available (Wielgoszewski et al., 2011a):

- Constant–current active–mode SThM (C-ASThM)
- Constant–temperature active–mode SThM (T-ASThM)

While operating the first of these, the amount of heat dissipated by the tip is constant. In the latter, it is modulated so that the temperature of the tip is constant. Both variants enable the measurement of the rate that the tip is cooled by the contact with the surface. In this way, the heat flow from the tip to the sample can be investigated, providing information on local thermal conductivity. Therefore, A-SThM can be also used to map the material homogeneity and to characterize novel materials to be further used in microelectronic and nanoelectronic technology, among other uses.

The tip can be heated directly or indirectly. The direct method utilizes current flowing through the tip, which in this case needs to be a resistive element (Fiege, Altes, Heiderhoff, & Balk, 1999; Vettiger et al., 2002; Lee & King, 2008; Dai, Corbin, & King, 2010; Janus et al., 2010; Wielgoszewski et al., 2010; Gomès et al., 2013; Kaźmierczak-Bałata et al., 2013). Indirect heating can be realized by using a separate heater affecting the temperature of a tip made of high-thermal-conductivity material (Brown, Hao, Cox, & Gallop, 2008). In both cases, a thermoresistive tip is expected to be used; however, there have been attempts to operate a laser-beam-heated thermocouple probe in A-SThM (Oesterschulze & Stopka, 1996) or to operate a Joule-heated thermocouple probe (Thiery, Gavignet, & Cretin, 2009).

In Figure 7, a schematic of a T-ASThM is shown. The temperature of the tip is measured as a difference between the measurement tip and the reference tip. In this specific setup, described more thoroughly by Wielgoszewski et al. (2011b), the signal from the current source comprises of two components: one for temperature measurement $f_{meas} \approx 10$ kHz and the other for tip heating $f_{heat} \in (0$ Hz, 1 kHz). The amplitude ratio would generally be in the range of $I_{heat} : I_{meas} \in (10, 30)$. The temperature signal is processed by a lock-in amplifier and is forwarded to a thermal proportional-integral-derivative (Th-PID) controller, the output of which is used to modulate the amplitude $I_{heat}$ of the heating signal, so that the tip temperature is kept constant. In this way, the Th-PID output signal carries information about the amount of heat needed to compensate for the cooling of the tip, resulting from its contact with the sample. These devices are presented in more detail in the section entitled "Measurement Devices for SThM," later in this chapter.

### 4.2.1 A-SThM in 3ω Variant

The basic idea of A-SThM is that in this mode the tip delivers heat to the sample in a controllable manner. An interesting improvement would be to use the third harmonic of the bridge output signal to allow the collection of

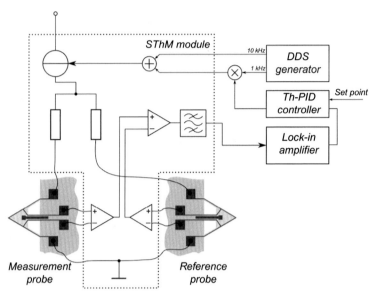

**Figure 7** Schematic of an active-mode SThM measurement setup in a constant tip temperature variant.

information on the sample thermal properties. Such a method has been known as a macroscopic one (Cahill, 1990). To apply it to A-SThM, the tip needs to be heated with alternating current (AC) at the fundamental frequency of $\omega$.

The instantaneous voltage drop at the SThM tip has a $3\omega$ component that depends on temperature change $\Delta T_{pr}$ and is independent of the average direct current (dc) tip resistance. However, the amplitude of this component is much less than of the lower harmonics, which means that a use of a precise lock-in amplifier (capable of processing the third harmonics), preceded by a preamplifier in the SThM module, is necessary (Lefèvre & Volz, 2005; Chirtoc & Henry, 2008; Wielgoszewski et al., 2011a). The $3\omega$-SThM is useful as a method of mapping the temperature and thermal conductivity (Altes, Heiderhoff, & Balk, 2002), but also for measuring the nanoscale thermal conductivity (Fiege et al., 1999; Ju, 1999; Huxtable et al., 2002).

## 4.3 DC and AC SThM Operation

Apart from choosing between passive and active mode, SThM users can decide on the way they supply the measurement circuit. The SThM probe (or the SThM measurement bridge) can be supplied via dc or AC. In most commercial SThM systems, simple dc voltage is used to enable probe resistance measurements. This allows basic passive- and active-mode operation. However, more advanced modes would use ac voltage: it is required for $3\omega$-SThM operation, but it is also useful in standard P-SThM and A-SThM, where it would help to obtain a better signal-to-noise ratio (Wielgoszewski et al., 2011b). Additionally, combined dc and AC may offer more stable operation with high-current heating of the sample (Bodzenta, Juszczyk, & Chirtoc, 2013). Last but not least, the interpretation of results in dc and AC SThM operation may be different, because in the latter, the influence of thermal waves must be taken into account (Gomès, Trannoy, Depasse, & Grossel, 2000, Gomès et al., 2001).

## 5. THERMAL PROBES FOR USE IN SThM

The development of SThM is closely related to the advances in design of thermal probes and complementary measurement devices. These allow for achieving better resolution and measurement accuracy. However, while improving the thermal probes, remember that one of the biggest breakthroughs in SThM was connected with the introduction of the simply designed and easy-to-use Wollaston probes. That is, the simpler the probe, the more impact on general development of the SThM it may have.

Of the papers on SThM that have been published since 2007, an increasing number of those reported experiments carried out using Kelvin Nanotechnology nanoprobes (Dobson et al., 2007; Kelvin Nanotechnology Ltd, 2015), which have been available commercially through vendors of SPM equipment for several years (Veeco Instruments, 2008; Anasys Instruments Corporation, 2010; Nanosurf, Inc, 2009). Also, it should be noted that probes with an implanted heater were made commercially available, which contributes to a significant increase in the number of nano-TA users (Veeco Instruments, 2008).

Advances in microelectronic and nanoelectronic technology enable the manufacture of probes of increasingly sophisticated design. One of the novelties is an attempt to establish multicantilever thermal sensing (Kim, Dai, & King, 2013). Moreover, it has been realized that SThM probes may be used not only as thermal sensors: resistive tips may also act as electrical sensors. By combining these two ideas, a multipurpose sensor is being developed (STREP NANOHEAT, 2012).

In Table 1, a comparison of SThM probe types is presented. The descriptions are provided in the following sections.

**Table 1** Comparison of SThM Probe Types

| Probe Type | Thermocouple | Thermoresistive | Bimetallic | Fluorescent |
|---|---|---|---|---|
| Thermal sensor | Chromel-alumel ThC Au-Cr ThC Pt-Cr ThC | Pt thermoresistor Pd thermoresistor Si thermoresistor | Al/Si bimetal | Er/Yb co-doped fluoride glass |
| Temperature change measurement principle | Electrical (Seebeck effect) | Electrical (resistance) | Frequency change | Optical |
| Passive mode | Available | Available | Available | Available |
| Active mode | Not available | Available | Not available | Not available |
| Liquid environment | Not available | Not available | Unknown | Available |
| Tip size | Usually <50 nm | 100 nm–5 $\mu$m | <20 nm | ~200 nm |
| Spatial resolution | Even below 20 nm | About 100 nm | Unknown | <1 $\mu$m |
| Durability | medium | high | Unknown | Unknown |
| Examples | Majumdar et al., 1995; Mills et al., 1998; Shi, Kwon, Wu, & Majumdar, 2000; Thiery et al., 2000; Kim et al., 2008, Kim, Jeong, Lee, & Reddy, 2012) | Dinwiddie et al., 1994; Edinger et al., 2001; Dobson et al., 2007; Wielgoszewski et al., 2010; Dai et al., 2010; Gaitas, Gianchandani, & Zhu, 2011; Menges, Riel, Stemmer, & Gotsmann, 2012; Tovee et al., 2012; Janus et al., 2014; Choi, Wu, & Lee, 2014; Hatakeyama et al., 2014; Hofer et al., 2015) | (Kim, Ono, & Esashi, 2009) | (Saïdi et al., 2009; Aigouy, Lalouat, Mortier, Löw, & Bergaud, 2011) |

*Note:* The resolution of temperature measurement is not included in this table, mostly because it does not depend only on the probe; in most cases it is estimated by the designers to be in the order of 0.1 K.

## 5.1 Thermocouple SThM Probes

Historically, the operation of the first SThM probes was based on the thermoelectric effect (Williams & Wickramasinghe, 1986b; Majumdar et al., 1993). The main element of these probes was a thin-film thermocouple that was placed as close to the measurement tip as possible. Using such a sensor, the information on tip temperature is achieved by recording the Seebeck-effect voltage, present at the junction.

In the newest solutions, the thermocouple SThM probes are fabricated in advanced microelectronic batch processes. These probes offer spatial resolution comparable to standard AFM probes and relatively good resolution of the temperature measurement. However, until recently, thermocouple probes have not been commercially available, although the development of the now-popular KNT-SThM resistive probe started with the thermocouple approach (Mills et al., 1998; Dobson et al., 2007). In 2013, Applied Nanostructures announced a new thermocouple probe, which, considering the data from the application note, may become another milestone in SThM development (Applied NanoStructures, 2013).

It is important that the voltage caused by the Seebeck effect is mostly related to the used materials. For this reason, temperature-related parameters of thermocouple probes are not very sensitive to elements such as the feature dimensions. The accompanying predictability and repeatability of these parameters is definitely a great advantage of these probes, along with their high resolution. The biggest disadvantage is their usability—they are basically only used for P-SThM. In Figure 8, one of the newer thermocouple probes is presented.

## 5.2 Thermoresistive SThM Probes

Among all the SThM probe types, the idea of a thermoresistive one is the simplest: such a probe utilizes a resistor as a thermal sensor. It can be realized as a metallic bent wire, metallic thin film, or a highly doped semiconductor

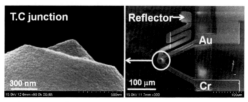

**Figure 8** A microfabricated thermocouple SThM probe with an Au–Cr junction. *Reprinted with permission from Kim et al. (2008), © by the AIP Publishing LLC.*

structure. For the metal-based designs, platinum is mostly used because of its high resistivity and high temperature coefficient of resistance (TCR). The other reason is that platinum has already been used in microelectronic integrated circuits, so its application does not require resolving additional technological problems. However, the newer batches of KNT-SThM probes have the sensing resistor made of palladium. The semiconductor probes are usually made of silicon.

The first thermoresisitive SThM probes were based on a Wollaston wire. Such a wire comprises of a platinum or platinum-rhodium core with diameter of 5 $\mu$m and a silver cladding with diameter of 70 $\mu$m, which was introduced by William Hyde Wollaston over 200 years ago (Wollaston, 1813). For use in SThM measurements, the Ag cladding is removed at the bending edge, so the surface is contacted by the Pt core itself (Dinwiddie et al., 1994). To enable optical feedback for AFM topography measurement, a mirror is placed on top of the wires (see Figure 9a). For their simple, almost macroscopic design, Wollaston-wire probes have often been used to investigate the thermal contact between the tip and the sample (Gomès et al., 2001; Lefèvre, Volz, Saulnier, Fuentes, & Trannoy, 2003, Lefèvre, Volz, & Chapuis, 2006). However, the disadvantages, which are low spatial resolution and complicated fabrication process, were significant enough to find a new way to utilize resistor-based probes in SThM measurements.

The development of thermoresistive probes was continued on two paths. One of them was to find ways to enhance spatial resolution while keeping the tip materially homogeneous. Good examples of this approach are a microfabricated cantilever with a fully platinum tip (Edinger et al., 2001; Rangelow, Gotszalk, Abedinov, Grabiec, & Edinger, 2001), a diamond-enhanced Wollaston probe (Brown et al., 2008) or a four-contact nanoprobe with an off-plane tip, designed by teams at Institute of Electron Technology in Warsaw and Wrocław University of Technology (ITE/WRUT nanoprobe; see Figure 9b) (Janus et al., 2010; Wielgoszewski et al., 2010). Another approach was to place a thin-film thermoresistor on an insulating support, the best example being the KNT-SThM nanoprobe made by Kelvin Nanotechnology (see Figure 9c), which is now made available at most suppliers of the SThM-enabled AFM systems (Kelvin Nanotechnology Ltd, 2015). An interesting four-terminal probe with a pyramidal tip has also been recently presented by Hatakeyama et al. (2014).

Another group of thermoresistive probes includes microcantilevers with doped-silicon tips and heaters. Such a design originates in the IBM's

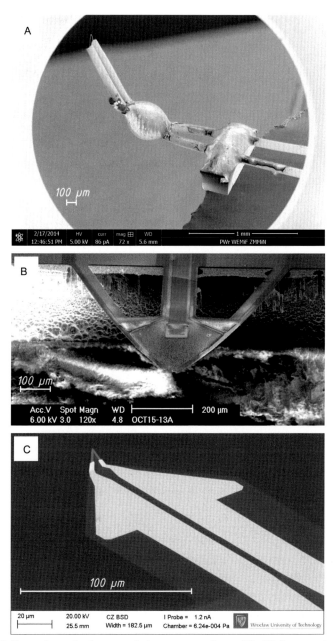

**Figure 9** Examples of thermoresistive nanoprobes: (A) Wollaston-wire probe, (B) ITE/WRUT nanoprobe, (C) KNT-SThM nanoprobe. Blue—platinum/palladium; light blue—silver; yellow—gold; green—Si₃N₄; pink—mounting glue; purple—glass. SEM images courtesy of Karolina Orłowska (Wrocław University of Technology), Yvonne Ritz (AMD Saxony), and Patrycja Szymczyk (Wrocław University of Technology). (See the color plate.)

*Millipede* device (Vettiger et al., 2002), although it was not originally meant to be used in SThM-like measurements. Semiconductor SThM probes are being developed in William P. King's group (Nelson & King, 2008), in variants with enhanced thermal insulation (Dai et al., 2010) and a platinum layer enabling thermoelectric measurement (Fletcher et al., 2012), and as matrices of few microcantilevers (Kim et al., 2013). Microcantilevers similar to those of the *Millipede* are used in A-SThM experiments carried out in IBM laboratories (Menges et al., 2012), while basic heated silicon probes are also available from Anasys Instruments, aimed at measuring thermal properties of polymers (Anasys Instruments Corporation, 2011; Asylum Research, 2011). Similar probes have also been developed by Choi et al. (2014).

One of the many advantages of thermoresistive probes is that in order to use them, no very specialized devices are needed. Simple dc-current Wheatstone bridges, which are used in commercial solutions, are accurate enough to provide basic thermal images. Nonetheless, the measurement resolution and accuracy can be improved by enhancing the probe design [e.g., by using four connections, as did Aubry et al. (2007); Janus et al. (2010, 2014)] or the measurement setup (Dobson et al., 2007; Wielgoszewski et al., 2011b; Bodzenta et al., 2013). Another important advantage is that thermoresistive probes can be utilized in both P-SThM and A-SThM modes, as the thermoresistor can be used easily as a heat source. Apart from thermal scanning, this kind of probe can be used for localized investigations of thermal properties like melting temperatures of polymers (Anasys Instruments Corporation, 2011). Recently, an SThM probe with a silicon heater was used to record surface topography based only on thermal interactions (Somnath & King, 2013).

## 5.3 Other Types of SThM Probes

Thermoelectric and thermoresistive SThM probes have a common weak point: while being used in contact-mode AFM, they need electric current to pass through. In case the investigated sample is electrically conductive, the obtained result may be significantly affected by an unexpected current distribution. The probe, having a certain electric potential, may change the state of the sample. Another area where the electric probe cannot be easily used is measuring in a liquid environment.

There are two probe types that can help to come over these limitations: bimetallic probes and fluorescent probes. Both are described in the next sections.

### 5.3.1 Bimetallic SThM Probes

The idea of using bimetallic microcantilevers can be traced back to 1994, when a silicon-aluminium microstructure was used in IBM laboratories to measure the temperature of the solution in which a chemical reaction occurred (Gimzewski, Gerber, Meyer, & Schlittler, 1994; Berger, Gerber, Gimzewski, Meyer, & Güntherodt, 1996). Similar probes were used to scan the surface in Majumdar's group in 1995 (Nakabeppu, Chandrachood, Wu, Lai, & Majumdar, 1995). In a more recent work, researchers fabricated a bimetallic subcantilever at the tip of a standard one using the focused ion beam (FIB; Kim et al., 2009). The authors stated that this enabled simultaneous imaging of the surface topography and temperature by detecting the deflection signal at two different frequencies. Such a solution may help in P-SThM investigations where contact mode in not suitable; however, the resolution may still need to be enhanced.

### 5.3.2 Fluorescent SThM Probes

Aigouy, Tessier, Mortier, and Charlot (2005) proposed that a tip with a fluorescent nanoparticle, made of erbium/ytterbium co-doped fluoride glass, also could be used as an SThM probe. They showed that the light intensity ratio for two photoluminescence peaks depends on temperature, so it can be used for temperature sensing. In later works, such a probe was used to map the temperature on metal lines with spatial resolution reaching 250 nm (Saïdi et al., 2009). They also managed to perform P-SThM measurements in a liquid environment with a resolution better than 1 $\mu$m (Aigouy et al., 2011).

## 6. MEASUREMENT DEVICES FOR SThM

The SThM module, mentioned at the beginning of the section entitled "SThM—The Development of the Technique," earlier in this chapter, can be realized in many ways. Specifically, the use of a fluorescent SThM probe would require a fully dedicated and unique measurement system, mostly because of including a specialized optical set-up. However, if the most popular thermoresistive and thermocouple SThM probes are considered, the SThM module would usually include the following:
- A current or voltage source (for thermoresistive probes)
- A measurement bridge (for thermoresistive probes)
- A preamplifier

Depending on the mode, the following external devices would be used:
- A function generator or a precision dc source
- A lock-in amplifier or a precision voltmeter

- A thermal proportional-integral-derivative (PID) controller for T-ASThM operation
- An absolute tip power/tip resistance meter for calibration purposes and quantitative A-SThM operation

Selected devices are characterized in the following subsections.

## 6.1 Current Source

The simplest way to use a measurement bridge would be to supply it from a voltage source—either a dc or an AC one. However, the use of a voltage-controlled current source should be considered as an alternative. A good choice to use in these procedures is an improved Howland pump (shown in Figure 10), which is a bipolar source with good performance in a large range of current amplitudes (Pease, 2008). It is especially useful if a two-component AC driving signal is used: in A-SThM, the heating current amplitude, $I_{heat}$, may be greater by two orders of magnitude than the measurement current amplitude, $I_{meas}$.

The SThM module may include signal conditioners to adjust the amplitude of controlling voltage signals, $V_{meas}$ and $V_{heat}$, which comprise the $V_{src}$ signal given o the source input. The resistor used for monitoring the current output should have tolerance of resistance no larger than $\pm 1\%$ and very good

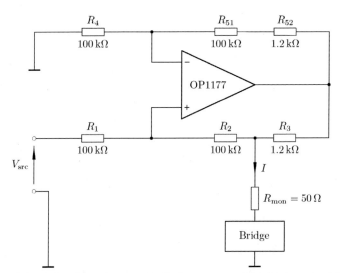

**Figure 10** Schematic of a voltage-controlled current source with improved Howland topology, included in SThM module used in experiments described by Wielgoszewski, Babij, Szeloch, & Gotszalk (2014). $V_{src}$—control voltage; $R_1$, $R_2$, $R_4$, $R_{51}$—precision-metallized resistors; $R_{mon}$—current monitor.

thermal stability (e.g., 25 ppm/K). The current monitoring signal should be collected using a precision instrumentation amplifier so that the measurement bridge circuit is not affected.

## 6.2 Measurement Bridges

Along with the microprobe, the measurement bridge is the most important part of the SThM measurement system. Since the introduction of thermoresistive SThM probes, the following four types of bridges were reported:
- Wheatstone bridge
- Transformer-isolated Wheatstone bridge
- Kelvin bridge
- Modified Wheatstone bridge for use with four-contact probes

The circuit schematics of these items are shown in Figure 11. Brief descriptions are also presented in the next sections.

### 6.2.1 Wheatstone Bridge

As the Wheatstone bridge (Figure 11a) was a widely applied solution for precise resistance measurement, well known for over a century (Christie, 1833; Wheatstone, 1843), it was not surprising that it was used in the first scanning thermal microscopes that utilized the Wollaston probe (Dinwiddie et al., 1994; Balk et al., 1997). It is still used in many commercial systems (Park Systems Corp, 2007, 2010; Anasys Instruments Corporation, 2010; Nanosurf, Inc, 2009). To operate such an SThM setup, only three resistors (and the probe) are needed. Generally, one of the resistors is a potentiometer used for pre-experiment bridge compensation. The disadvantage is that it does not take advantage of a four-contact probe, if one is used.

### 6.2.2 Transformer-Isolated Wheatstone Bridge

The use of an electrical (i.e., resistive) probe with a regular bridge requires that the probe tip have nonzero electric potential. In case of in situ measurements of microelectronic and nanoelectronic devices, this fact may cause disturbances in the system, both on the probe and sample sides. In some cases, it may even end up destroying an electrostatically sensitive nanoprobe (Dobson et al., 2007). Therefore, a transformer-isolated measurement bridge was proposed (Figure 11b), which helped to minimize the electrostatic force between the tip and the sample (Dobson et al., 2007). It has to be noted that for such an operation, an AC signal is required, while in general, the thermal probes may behave differently at various frequencies; e.g., in a KNT-SThM probe at a frequency of 1 MHz, a capacitive behavior

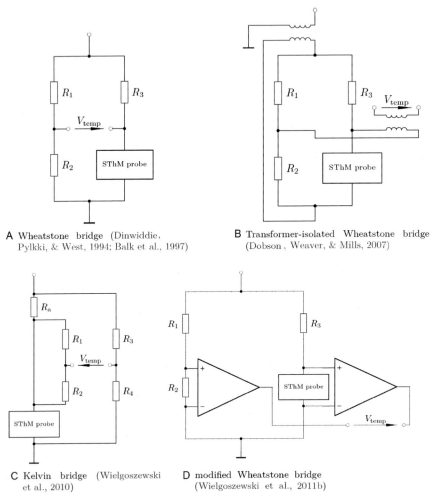

**A** Wheatstone bridge (Dinwiddie, Pylkki, & West, 1994; Balk et al., 1997)

**B** Transformer-isolated Wheatstone bridge (Dobson , Weaver, & Mills, 2007)

**C** Kelvin bridge (Wielgoszewski et al., 2010)

**D** modified Wheatstone bridge (Wielgoszewski et al., 2011b)

**Figure 11** Schematics of SThM measurement bridges. Bridges in panels (c) and (d) are designed for four-terminal SThM probes.

of the system Au lead–$Si_3N_4$ layer–substrate becomes significant, which causes leakage and may affect SThM measurement.

### 6.2.3 Kelvin Bridge

The Kelvin bridge was originally designed for measurements of very low resistance (Thomson, 1862). To benefit from this, a four-point contact to the measured resistor is required. In terms of SThM use, such a topology was suggested along with a four-terminal probe design by Janus et al. (2010).

The general idea is that the most current flows through the main branch of the circuit (the $R_a$ resistor and the probe in Figure 11c). For this reason, the branch with the bridge is built of resistors with resistance of 1 M$\Omega$ or more. However, although such a topology may be useful for four-terminal SThM probes, the sensitivity of an SThM Kelvin bridge may not be sufficient, while the noise level will be significantly increased (Wielgoszewski et al., 2011b).

### 6.2.4 Modified Wheatstone Bridge

A better solution for a four-terminal probe is a simple bridge in which the regular voltage difference measurement is replaced by measurement of the difference of the amplified voltages at the probe voltage terminals (Figure 11d). Provided suitable probes and precision operational amplifiers are used, better accuracy should be achieved. Moreover, such a setup revealed to have greater measurement sensitivity than the standard Wheatstone and Kelvin bridges [Figure 12, Wielgoszewski et al. (2011b)].

## 6.3 Preamplifier

All the bridges described in the previous section have a single voltage output, from which information on temperature changes can be recorded. The same kind of signal would be used for thermocouple probes. In general, it would be better in terms of noise handling to have the signal preamplified before it is passed to the lock-in amplifier. Such a device mostly should include a

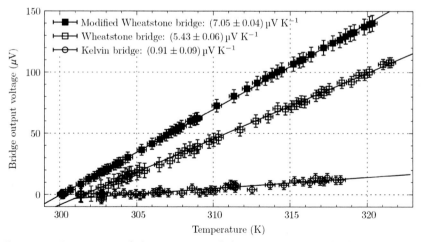

**Figure 12** Comparison of the sensitivity of three measurement bridge topologies (Wielgoszewski et al., 2011b). The experiments were carried out in as similar conditions as possible, including constant probe current $I_{pr}$.

precision instrumentation amplifier; however, in the case of AC–current operation, the use of band–pass filters is also encouraged.

## 6.4 Thermal PID Controller

As was mentioned in the section entitled "A-SThM," later in this chapter, it is possible to operate an SThM system in constant tip temperature mode (T-ASThM). To enable it, a thermal PID controller (Th-PID) is required to control the tip temperature. The processed temperature signal (i.e., the output of the lock–in amplifier that handles the bridge output $V_{temp}$) should be passed to the Th-PID input, while the Th-PID output should be modulating the amplitude of the heating current $I_{heat}$. The user can control the $V_{temp}$ setpoint, which allows setting of the tip temperature (see also Figure 7). Once a constant tip temperature is ensured, quantitative conclusions on heat flow between the tip and the sample can be drawn.

## 6.5 Tip Power and Tip Resistance Meter

To facilitate quantitative measurements, analog processing of absolute tip voltage $V_{pr}$ and tip current $I_{pr}$ signals may be useful. If the SThM module provides such signals, they can be multiplied or divided to obtain instantaneous tip power $P_{pr}$ and tip resistance $R_{pr}$. Regardless of the fact if AC or dc excitation was used, the tip power and tip resistance meter (i.e., a P/R meter) would provide dc signals, which are also suitable for direct recording in the AFM system during scanning of the surface. Such a device allows relatively easy calibration of the $P_{pr}$ and $R_{pr}$ outputs by using a set of reference resistors instead of a probe, therefore is also useful for calibration of the SThM system (Wielgoszewski, Babij, Szeloch, & Gotszalk, 2014). A general view of a SThM system is presented in Figure 13.

## 7. METROLOGY IN SThM

Since the first SThM system was built, it has been possible to record qualitative temperature maps and images of thermal properties. It is still useful as such; however, the number of possible applications of quantitative SThM is increasing. Sometimes information on the relative temperature is enough, as well as information on thermal conductivity contrast. Nonetheless, if reliable, fully quantitative measurements are available, they allow investigation of fundamental phenomena related to heat flow, and the conclusions also are far more interesting. Several recent studies of graphene are

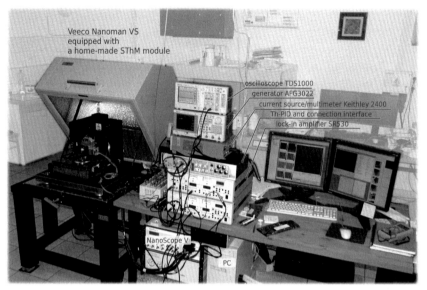

**Figure 13** An overview of a commercial AFM equipped with a homemade SThM system, as used in the Nanometrology Group at Wrocław University of Technology. (See the color plate.)

very good examples of this (Pumarol et al., 2012; Menges, Riel, Stemmer, Dimitrakopoulos, & Gotsmann, 2013; Yoon et al., 2014).

The fact that the need of a reliable quantitative SThM is currently an important challenge is confirmed by two EU projects being run: NANOHEAT (STREP NANOHEAT, 2012) and QUANTIHEAT (IP QUANTIHEAT, 2013).

## 7.1 Metrology in Scanning Probe Microscopy

Generally, traceable measurements are a challenge all over the scanning probe microscopy methods. To consider a simple topography measurement, the metrology chain consists of at least four links (Klapetek, 2013):

1. National standard of length—A wavelength-stabilized laser
2. An interferometer in a metrological AFM, calibrated using the national standard of length
3. Calibration grating for AFMs, calibrated using a metrological AFM
4. A user-owned AFM system calibrated using a calibration grating

It must be noted that there are not many metrological AFMs: Yacoot and Koenders (2011) listed 27 of them, built since the 1990s around the world. As such systems are not widely available, certification of calibration gratings is time-consuming and relatively expensive, and regular AFM users do not pay enough attention to traceable measurements (although they usually use basic, noncertified gratings).

Calibrated topography measurement is only the beginning. Moving on to advanced SPM techniques, we find that calibration setups or samples are definitely not common, especially if standard traceability is concerned.

## 7.2 Calibration in SThM

The term *calibration* means to establish a relation between the indication of a measurement device and the value of a physical quantity represented by an etalon (JCGM 200:2012, 2012). Currently, the temperature reference is the International Temperature Scale of 1990 (ITS-90), described in Preston-Thomas (1990). It consists of 17 defined fixed temperature points within the range of 0.65–1358 K, which are mostly defined as triple points and freezing points of certain substances, accompanied by interpolation equations for the calibration curves.

A standard-based calibration method for SThM should generally provide a variable-temperature contact point for an SThM tip, which could be kept in a known temperature. The reference temperature may be provided by an instrument such as a Pt100 resistor; however, it would be better to use few of the fixed points included in the ITS-90. As can be seen in Table 2, there is only one method to date that tries to fulfill this requirement (Wielgoszewski, Babij, Szeloch, & Gotszalk, 2014).

Among the methods listed in Table 2, two in particular should be taken into consideration: one method involving the microelectronic hot-plate temperature standard (Dobson et al., 2005) and another method involving the use of ITS-90 materials (Wielgoszewski, Babij, Szeloch, & Gotszalk, 2014). The first of these is a microfabricated Johnson thermal noise reference; therefore, the calibration is based directly on the thermodynamic temperature scale. The thermal noise of a reference $Ni_{0.6}Cr_{0.4}$ resistor is used to calibrate an Au–Pd thermocouple, which is later used for SThM probe contact-based calibration. The other method uses melting points of gallium and benzophenone: in this procedure, the SThM tip is brought into contact with a reference material, which is then melted. The cantilever deflection signal is recorded simultaneously with the tip temperature signal so that the moment of melting can be spotted and the temperature linked to the tip temperature voltage signal or to the absolute tip resistance. Accuracy of the two methods is reported to be in the order of 1 K, which makes them promising in terms of further development of SThM metrology.

**Table 2** Calibration Methods Used in SThM

| SThM Probe Type | Calibration Method | References |
|---|---|---|
| Thermoresistive (temperature) | Microelectronic hot-plate stage | Aubry et al. (2007); Barbosa and Slifka (2008); Zhang, Dobson, and Weaver (2012) |
| | Macroscopic hot plate | Wielgoszewski et al. (2011b); Menges et al. (2012) |
| | Raman spectroscopy of a hot structure | Yu et al. (2011); Menges et al. (2012) |
| | Immersion in PDMS of known temperature | Buzin, Kamasa, Pyda, and Wunderlich (2002) |
| | Melting point of polymers | Lee and Gianchandani (2004) |
| | Contact-based thermocouple reference measurement | Gomès et al. (2001) |
| | Johnson noise (of a resistor in microfabricated stage or a gold wire) | Dobson et al. (2007); Gaitas, Wolgast, Covington, and Kurdak (2013) |
| | Melting point of ITS-90 reference materials | Wielgoszewski, Babij, Szeloch, and Gotszalk (2014) |
| Thermoresistive (thermal conductivity) | Vertical silicon nanowires | Puyoo, Grauby, Rampnoux, Rouviere, and Dilhaire (2011) |
| | Materials with known conductivity | Ruiz, Sun, Pollak, and Venkatraman (1998); Guo, Trannoy, and Lu (2006); Bodzenta, Kaźmierczak-Bałata, Lorenc, and Juszczyk (2010) |
| | No information | Price, Reading, Hammiche, Pollock, and Branch (1999); Tovee et al. (2012) |
| Thermocouple | Macroscopic stage with a Peltier module | Luo, Majumdar, Herrick, and Petroff (1997); Genix, Vairac, and Cretin (2009) |
| | Microelectronic stage with a heater | Majumdar et al. (1995); Shi et al. (2001) |
| | Contact-based thermocouple reference measurement | Shi et al. (2000) |

*Continued*

**Table 2** Calibration Methods Used in SThM—cont'd
**SThM Probe**

| Type | Calibration Method | References |
|------|-------------------|-----------|
| | No calibration performed (known Seebeck coefficient for K-type thermocouple was used) | Majumdar et al. (1993) |
| Fluorescent | No calibration performed (analytical relationship of intensity ratio and temperature was used) | Samson et al. (2007); Saïdi et al. (2009); Aigouy et al. (2011) |

## 7.3 TThM: Tip Shape Impact on SThM Image

It has already been mentioned that the measurement tip, or even the probe as a whole, determine what can be seen using an SPM system. One of many aspects of this relationship is the tip shape, which always affects the obtained image, whether only the topography was recorded or advanced modes were also involved. To begin with, in the 1990s, John Villarrubia proposed a blind-tip reconstruction method (Villarrubia, 1994, 1997). These were later enhanced by Tian, Qian, and Villarrubia (2008) and Jóźwiak et al. (2012) and now are known as *regularized blind-tip reconstruction (RBTR)*. The general idea behind the RBTR method is to find the shape of the largest possible tip that could have been used to produce the analyzed topography image.

Because the details on tip–sample thermal contact play a crucial role in the analysis of any SThM results, many approaches to the analysis of phenomena occurring between the probe and the sample surface have been proposed, all of them being a combination of experiment, analytical analysis, and simulation (Luo, Shi, Varesi, & Majumdar, 1997; Gomès et al., 2001; Lefèvre et al., 2006; Thiery et al., 2008; Bodzenta et al., 2010; Tovee et al., 2012; Gotsmann & Lantz, 2013; Kim et al., 2014). Moreover, all of these works included some assumptions about the tip shape, which often turned out to have a significant impact on the conclusions.

RBTR method and the tip–sample issues in SThM have been combined in the tip thermal mapping (TThM) method (Jóźwiak et al., 2013). To have a tip characterized using TThM, a T-ASThM scan of a sample with homogeneous thermal conductivity is needed. Once the images are recorded, the

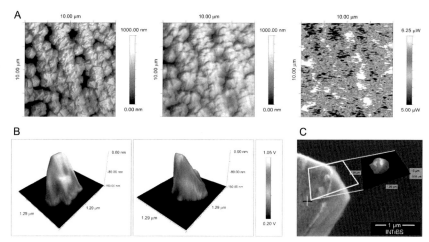

**Figure 14** (A) Images recorded for further processing in TThM: the reference high-resolution AFM topography (left) and simultaneously recorded SThM topography and T-ASThM heat dissipation from the tip to the sample; (B) the estimated tip shape with the thermal map seen as color overlay; (C) SEM image as a confirmation of the correct tip shape estimation. High-resolution SEM image courtesy of Leszek Kępiński (Institute of Low Temperatures and Structure Research of the Polish Academy of Sciences); blue—platinum; green—SiO₂. *Reprinted from Jóźwiak et al. (2013), with permission from Elsevier.* (See the color plate.)

SThM tip shape is estimated from the SThM topography image. Afterward, the area where the tip contacts the surface is calculated, while another high-resolution standard–AFM topography image of the surface is used as a reference (see Figure 14a). Based on the SThM heat dissipation image, it can be calculated how much heat flows through each point of the tip, and a thermal map of the tip can be obtained. The method is especially useful in the case of SThM probes with a nonhomogeneous tip, like KNT-SThM: in Figures 14b and 14c, it can be noticed that there was more heat flow on the side of the thin-film resistor (Jóźwiak et al., 2013).

## 8. SThM RESULTS

The reader may have already noticed that there are plenty of possible applications of various SThM modes and variants. To conclude this chapter, a small amount of the reported results is described to give a general idea of what the images taken using SThM can be like.

## 8.1 Passive-Mode SThM

In Figure 15, a typical P–SThM image is shown. Topography of the sample, which is a polycrystalline-silicon microfuse, fabricated in Institute of Electron Technology in Warsaw, Poland, is overlaid with the temperature map of the surface. The dimensions of the fuse are: (i) length of 35 $\mu$m, (ii) base width of 8 $\mu$m, and (iii) peak height of 0.5 $\mu$m. During the scan, the fuse was self-heating as a result of 10 mA of current, which meant a current density of about $5.5 \times 10^5$ A cm$^{-2}$. It can be seen that apparently the fuse is working properly, because the highest temperature point can be found approximately in the center of the fuse. A KNT-SThM and a homemade SThM module was used for this experiment (Wielgoszewski et al., 2014).

## 8.2 Active-Mode SThM

As described in the section entitled "A–SThM," earlier in this chapter, A–SThM is generally used for investigations of thermal conductivity or heat dissipation. Menges et al. (2013) carried out a study of thermal resistance of graphene layers, a result of which is shown in Figure 16. SThM was operated in active mode, enabling recognition of single-layer and bilayer graphene, which was formed on a SiC substrate. A *Millipede*-like thermal probe in a vacuum-operated, homemade SThM system was used to obtain these images.

If quantitative T–ASThM is employed, analysis of the heat flow may lead to more than imaging and recognition of features. Wielgoszewski et al.

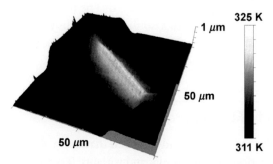

**Figure 15** Passive-mode SThM image of a microfuse made of polycrystalline silicon, obtained using a calibrated tip while a 10-mA current was flowing through the structure; the temperature map is overlaid on the recorded surface topography. *Reprinted from Wielgoszewski et al. (2014), with permission from Elsevier.* (See the color plate.)

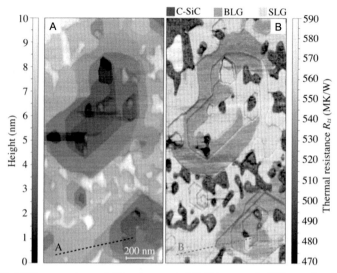

**Figure 16** (A) Topography and (B) active-mode SThM image of a SiC sample with single-layer graphene (SLG) and bilayer graphene (BLG) formed on the surface. *Reprinted with permission from Menges et al. (2013) © by the American Physical Society.* (See the color plate.)

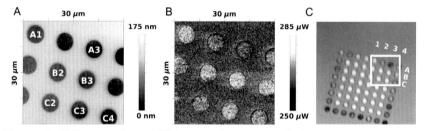

**Figure 17** (A) Topography and (B) tip power images recorded in constant-temperature active-mode SThM, accompanied by (C) optical image of the structure. *Reprinted from Wielgoszewski et al. (2014), with permission from Elsevier.* (See the color plate.)

(2014) investigated through-silicon vias (TSVs), which enabled the estimation of thermal conductivity of the antidiffusion barrier, which is used to prevent the migration of copper into the substrate material. Results of T-ASThM measurements of the TSV matrix is shown in Figure 17. The calculation results, with data input being only the T-ASThM data for each of the pixels, showed the difference in tip–sample contact thermal resistance for two groups of TSVs: {A1, B2, B3} and {A3, C2, C3, C4}.

## 9. FINAL REMARKS

This chapter has introduced the SThM, a specialized scanning probe microscopy method. Basic modes have been described, as well as the SThM probe types and few of the possible specific amendments. As this work is just a basic introduction, the reader is encouraged to consult the references, all of them describing strikingly interesting investigations carried out using various SThM variants. If it is considered that SThM has already achieved a resolution that is on the order of magnitude of mean-free-paths of phonons in graphene, an increase in interest in nanoscale thermal investigations, obviously not limited to carbon-related materials, can be expected. As was mentioned, many improvements to the SThM has only recently been introduced or proposed. Therefore, it seems that the best is still to come—and we look forward to it.

### ACKNOWLEDGMENTS

The authors would like to thank all the fantastic people with whom they have collaborated on SThM-related projects, and all members of the Student Scientific Association SPENT of the Faculty of Microsystem Electronics and Photonics at Wrocław University of Technology, without whom the research would not be so enjoyable. The work was partially supported by the Foundation for Polish Science within the *Mistrz* (Master) grant no. 4/2012.

### REFERENCES

Abraham, D. W., Williams, C. C., Slinkman, J., & Wickramasinghe, H. K. (1991). Lateral dopant profiling in semiconductors by force microscopy using capacitive detection. *Journal of Vacuum Science & Technology B: Microelectronics and Nanometer Structures, 9*(2), 703. http://dx.doi.org/10.1116/1.585536.
Aigouy, L., Lalouat, L., Mortier, M., Löw, P., & Bergaud, C. (2011). Note: A scanning thermal probe microscope that operates in liquids. *Review of Scientific Instruments, 82*(3), 036106. http://dx.doi.org/10.1063/1.3567794.
Aigouy, L., Tessier, G., Mortier, M., & Charlot, B. (2005). Scanning thermal imaging of microelectronic circuits with a fluorescent nanoprobe. *Applied Physics Letters, 87*(18), 184105. http://dx.doi.org/10.1063/1.2123384.
Altes, A., Heiderhoff, R., & Balk, L. J. (2002). Complementary surface investigation of diamond by scanning thermal microscopy and scanning near-field cathodoluminescence. *International Journal of Modern Physics B, 16*(06n07), 895–899. http://dx.doi.org/10.1142/S0217979202010579.
Anasys Instruments Corporation. (2010). *Scanning thermal microscopy.* Application note, http://www.anasysinstruments.com/SThM.php.
Anasys Instruments Corporation. (2011). *Therma-Lever™ probes.* Application note, www.anasysinstruments.com/nano-TAprobes.pdf.
Applied NanoStructures. (2013). *VertiSense™—Scanning thermal microscopy module and probes.* Application note, http://www.appnano.com/static/pub/documents/Vertisense_Thermal.pdf.

Asylum Research. (2011). *Ztherm™ modulated thermal analysis option for asylum research AFMs*. Application note, http://www.asylumresearch.com/Products/Ztherm/Ztherm DSHR.pdf.

Aubry, R., Jacquet, J.-C., Weaver, J., Durand, O., Dobson, P., Mills, G., et al. (2007). SThM temperature mapping and nonlinear thermal resistance evolution with bias on AlGaN/GaN HEMT devices. *IEEE. Transactions on Electron Devices, 54*(3), 385–390. http://dx.doi.org/10.1109/TED.2006.890380.

Avila, A., & Bhushan, B. (2010). Electrical measurement techniques in atomic force microscopy. *Critical Reviews in Solid State and Materials Sciences, 35*(1), 38–51. http://dx.doi.org/10.1080/10408430903362230.

Balk, L., Cramer, R., & Fiege, G. (1997). Thermal analyses by means of scanning probe microscopy. *Proceedings of the 1997 6th International Symposium on the Physical and Failure Analysis of Integrated Circuits, Singapore: IEEE*. http://dx.doi.org/10.1109/IPFA.1997.638063.

Barbosa, N., & Slifka, A. J. (2008). Spatially and temporally resolved thermal imaging of cyclically heated interconnects by use of scanning thermal microscopy. *Microscopy Research and Technique, 71*(8), 579–584. http://dx.doi.org/10.1002/jemt.20589.

Berger, R., Gerber, C., Gimzewski, J. K., Meyer, E., & Güntherodt, H. J. (1996). Thermal analysis using a micromechanical calorimeter. *Applied Physics Letters, 69*(1), 40. http://dx.doi.org/10.1063/1.118111.

Bhushan, B., & Goldade, A. V. (2000). Measurements and analysis of surface potential change during wear of single-crystal silicon (100) at ultralow loads using Kelvin probe microscopy. *Applied Surface Science, 157*(4), 373–381. http://dx.doi.org/10.1016/S0169-4332(99)00553-X.

Binnig, G., Quate, C. F., & Gerber, C. (1986). Atomic force microscope. *Physical Review Letters, 56*(9), 930–933. http://dx.doi.org/10.1103/PhysRevLett.56.930.

Binnig, G., & Rohrer, H. (1987). Scanning tunneling microscopy—From birth to adolescence. *Reviews of Modern Physics, 59*(3), 615–625. http://dx.doi.org/10.1103/RevModPhys.59.615.

Binnig, G., Rohrer, H., Gerber, C., & Weibel, E. (1982). Surface studies by scanning tunneling microscopy. *Physical Review Letters, 49*(1), 57–61. http://dx.doi.org/10.1103/PhysRevLett.49.57.

Bodzenta, J., Juszczyk, J., & Chirtoc, M. (2013). Quantitative scanning thermal microscopy based on determination of thermal probe dynamic resistance. *Review of Scientific Instruments, 84*(9), 093702. http://dx.doi.org/10.1063/1.4819738.

Bodzenta, J., Kaźmierczak-Bałata, A., Lorenc, M., & Juszczyk, J. (2010). Analysis of possibilities of application of nanofabricated thermal probes to quantitative thermal measurements. *International Journal of Thermophysics, 31*(1), 150–162. http://dx.doi.org/10.1007/s10765-009-0659-2.

Braun, T., Ghatkesar, M. K., Backmann, N., Grange, W., Boulanger, P., Letellier, L., et al. (2009). Quantitative time-resolved measurement of membrane protein–ligand interactions using microcantilever array sensors. *Nature Nanotechnology, 4*(3), 179–185. http://dx.doi.org/10.1038/nnano.2008.398.

Brown, E., Hao, L., Cox, D. C., & Gallop, J. C. (2008). Scanning thermal microscopy probe capable of simultaneous electrical imaging and the addition of diamond tip. *Journal of Physics: Conference Series, 100*(5), 052012. http://dx.doi.org/10.1088/1742-6596/100/5/052012.

Buzin, A., Kamasa, P., Pyda, M., & Wunderlich, B. (2002). Application of a Wollaston wire probe for quantitative thermal analysis. *Thermochimica Acta, 381*(1), 9–18. http://dx.doi.org/10.1016/S0040-6031(01)00648-7.

Cahill, D. G. (1990). Thermal conductivity measurement from 30 to 750 K: The 3ω method. *Review of Scientific Instruments, 61*(2), 802. http://dx.doi.org/10.1063/1.1141498.

Chirtoc, M., & Henry, J. F. (2008). $3\omega$ hot wire method for micro-heat transfer measurements: From anemometry to scanning thermal microscopy (SThM). *European Physical Journal Special Topics*, *153*(1), 343–348. http://dx.doi.org/10.1140/epjst/e2008-00458-8.

Choi, Y.-S., Wu, X., & Lee, D.-W. (2014). Selective nano-patterning of graphene using a heated atomic force microscope tip. *Review of Scientific Instruments*, *85*(4), 045002. http://dx.doi.org/10.1063/1.4870588.

Christie, S. H. (1833). Experimental determination of the laws of magneto-electric induction in different masses of the same metal, and of its intensity in different metals. *Philosophical Transactions of the Royal Society of London*, *123*, 95–142. http://dx.doi.org/10.1098/rstl.1833.0011.

Dai, Z., Corbin, E. A., & King, W. P. (2010). A microcantilever heater-thermometer with a thermal isolation layer for making thermal nanotopography measurements. *Nanotechnology*, *21*(5), 055503. http://dx.doi.org/10.1088/0957-4484/21/5/055503.

De Wolf, P., Snauwaert, J., Clarysse, T., Vandervorst, W., & Hellemans, L. (1995). Characterization of a point-contact on silicon using force microscopy-supported resistance measurements. *Applied Physics Letters*, *66*(12), 1530. http://dx.doi.org/10.1063/1.113636.

Dinwiddie, R. B., Pylkki, R. J., & West, P. E. (1994). Thermal conductivity contrast imaging with a scanning thermal microscope. In T. W. Tong (Ed.), *Thermal conductivity 22* (pp. 668–677). Lancaster, PA: Technomic.

Dobson, P. S., Mills, G., & Weaver, J. M. R. (2005). Microfabricated temperature standard based on Johnson noise measurement for the calibration of micro- and nano-thermometers. *Review of Scientific Instruments*, *76*(5), 054901. http://dx.doi.org/10.1063/1.1899463.

Dobson, P. S., Weaver, J. M. R., & Mills, G. (2007). New methods for calibrated scanning thermal microscopy (SThM) (pp. 708–711). 2007 IEEE Sensors, Atlanta, GA: IEEE; http://dx.doi.org/10.1109/ICSENS.2007.4388498.

Edinger, K., Gotszalk, T., & Rangelow, I. W. (2001). Novel high resolution scanning thermal probe. *Journal of Vacuum Science & Technology B: Microelectronics and Nanometer Structures*, *19*(6), 2856. http://dx.doi.org/10.1116/1.1420580.

Fiege, G., Altes, A., Heiderhoff, R., & Balk, L. (1999). Quantitative thermal conductivity measurements with nanometre resolution. *Journal of Physics D: Applied Physics*, *32*, L13–L17. http://dx.doi.org/10.1088/0022-3727/32/5/003.

Fletcher, P. C., Lee, B., & King, W. P. (2012). Thermoelectric voltage at a nanometer-scale heated tip point contact. *Nanotechnology*, *23*(3), 035401. http://dx.doi.org/10.1088/0957-4484/23/3/035401.

Gaitas, A., Gianchandani, S., & Zhu, W. (2011). A piezo-thermal probe for thermomechanical analysis. *Review of Scientific Instruments*, *82*, 053701. http://dx.doi.org/10.1063/1.3587624.

Gaitas, A., Wolgast, S., Covington, E., & Kurdak, C. (2013). Hot-spot detection and calibration of a scanning thermal probe with a noise thermometry gold wire sample. *Journal of Applied Physics*, *113*, 074304. http://dx.doi.org/10.1063/1.4792656.

Gajewski, K., Kopiec, D., Moczała, M., Piotrowicz, A., Zielony, M., Wielgoszewski, G., et al. (2015). Scanning probe microscopy investigations of the electrical properties of chemical vapor deposited graphene grown on a 6H-SiC substrate. *Micron*, *68*(Jan), 17–22. http://dx.doi.org/10.1016/j.micron.2014.08.005.

Genix, M., Vairac, P., & Cretin, B. (2009). Local temperature surface measurement with intrinsic thermocouple. *International Journal of Thermal Sciences*, *48*(9), 1679–1682. http://dx.doi.org/10.1016/j.ijthermalsci.2009.01.020.

Gimzewski, J., Gerber, C., Meyer, E., & Schlittler, R. (1994). Observation of a chemical reaction using a micromechanical sensor. *Chemical Physics Letters*, *217*(5–6), 589–594. http://dx.doi.org/10.1016/0009-2614(93)E1419-H.

Girard, P. (2001). Electrostatic force microscopy: Principles and some applications to semiconductors. *Nanotechnology*, *12*(4), 485–490. http://dx.doi.org/10.1088/0957-4484/12/4/321.

Gomès, S., Newby, P., Canut, B., Termentzidis, K., Marty, O., Fréchette, L., et al. (2013). Characterization of the thermal conductivity of insulating thin films by scanning thermal microscopy. *Microelectronics Journal*, *44*(11), 1029–1034. http://dx.doi.org/10.1016/j.mejo.2012.07.006.

Gomès, S., Trannoy, N., Depasse, F., & Grossel, P. (2000). A.C. scanning thermal microscopy: Tip-sample interaction and buried defects modellings. *International Journal of Thermal Sciences*, *39*(4), 526–531. http://dx.doi.org/10.1016/S1290-0729(00)00232-5.

Gomès, S., Trannoy, N., Grossel, P., Depasse, F., Bainier, C., & Charraut, D. (2001). D.C. scanning thermal microscopy: Characterisation and interpretation of the measurement. *International Journal of Thermal Sciences*, *40*(11), 949–958. http://dx.doi.org/10.1016/S1290-0729(01)01281-9.

Gotsmann, B., & Lantz, M. A. (Jan. 2013). Quantized thermal transport across contacts of rough surfaces. *Nature Materials*, *12*(1), 59–65. http://dx.doi.org/10.1038/nmat3460.

Gotszalk, T. P. (2004). *Systemy mikroskopii bliskich oddziaływań w badaniach mikro- i nanostruktur (Scanning probe microscopy systems in investigations of micro- and nanostructures)*. Wrocław: Oficyna Wydawnicza Politechniki Wrocławskiej.

Gotszalk, T., Grabiec, P., & Rangelow, I. W. (2000). Piezoresistive sensors for scanning probe microscopy. *Ultramicroscopy*, *82*(1–4), 39–48. http://dx.doi.org/10.1016/S0304-3991(99)00171-0.

Guo, F., Trannoy, N., & Lu, J. (2006). Characterization of the thermal properties by scanning thermal microscopy in ultrafine-grained iron surface layer produced by ultrasonic shot peening. *Materials Chemistry and Physics*, *96*(1), 59–65. http://dx.doi.org/10.1016/j.matchemphys.2005.05.055.

Güthner, P., & Dransfeld, K. (1992). Local poling of ferroelectric polymers by scanning force microscopy. *Applied Physics Letters*, *61*(9), 1137. http://dx.doi.org/10.1063/1.107693.

Hatakeyama, K., Sarajlic, E., Siekman, M. H., Jalabert, L., Fujita, H., Tas, N., et al. (2014). *Wafer-scale fabrication of scanning thermal probes with integrated metal nanowire resistive elements for sensing and heating*. In 2014 IEEE 27th International Conference on Micro Electro Mechanical Systems (MEMS) (pp. 1111–1114). http://dx.doi.org/10.1109/MEMSYS.2014.6765840.

Haugstad, G. (2012). *Atomic force microscopy: Understanding basic modes and advanced applications*. Hoboken, NJ: John Wiley & Sons, Inc. http://dx.doi.org/10.1002/9781118360668.

Hembacher, S., Giessibl, F. J., & Mannhart, J. (2004). Force microscopy with light-atom probes. *Science*, *305*(5682), 380–383. http://dx.doi.org/10.1126/science.1099730.

Hofer, M., Ivanov, T., Rudek, M., Kopiec, D., Guliyev, E., Gotszalk, T. P., & Rangelow, I. W. (2015). Fabrication of self-actuated piezoresistive thermal probes. *Microelectronic Engineering*, *145*, 32–37. http://dx.doi.org/10.1016/j.mee.2015.02.016.

Huey, B. D., Ramanujan, C., Bobji, M., Blendell, J., White, G., Szoszkiewicz, R., et al. (2004). The importance of distributed loading and cantilever angle in piezo-force microscopy. *Journal of Electroceramics*, *13*(1–3), 287–291. http://dx.doi.org/10.1007/s10832-004-5114-y.

Huxtable, S. T., Abramson, A. R., Tien, C.-L., Majumdar, A., LaBounty, C., Fan, X., et al. (2002). Thermal conductivity of Si/SiGe and SiGe/SiGe superlattices. *Applied Physics Letters*, *80*(10), 1737. http://dx.doi.org/10.1063/1.1455693.

Veeco Instruments. (2008). Veeco Instruments thermal analysis nanoscale material characterization/identification by AFM. Application note.

IP QUANTIHEAT. (2013). Quantitative scanning probe microscopy techniques for heat transfer management in nanomaterials and nanodevices. Project Within the 7th Framework Programme of the European Union. http://www.quantiheat.org.

Ivanova, K., Sarov, Y., Ivanov, T., Frank, A., Zöllner, J., Bitterlich, C., et al. (2008). Scanning proximal probes for parallel imaging and lithography. *Journal of Vacuum Science & Technology B: Microelectronics and Nanometer Structures, 26*(6), 2367–2373. http://dx.doi. org/10.1116/1.2990789.

Janus, P., Grabiec, P., Sierakowski, A., Gotszalk, T., Rudek, M., Kopiec, D., et al. (2014). *Design, technology, and application of integrated piezoresistive scanning thermal microscopy (SThM) microcantilever*. In M. T. Postek, D. E. Newbury, S. F. Platek, & T. K. Maugel (Eds.), *Proceedings of the SPIE 9236: Scanning Microscopies 2014, 92360R*. http://dx.doi.org/10.1117/12.2066240.

Janus, P., Szmigiel, D., Weisheit, M., Wielgoszewski, G., Ritz, Y., Grabiec, P., et al. (2010). Novel SThM nanoprobe for thermal properties investigation of micro- and nanoelectronic devices. *Microelectronic Engineering, 87*(5–8), 1370–1374. http://dx.doi. org/10.1016/j.mee.2009.11.178.

JCGM 200:2012. (2012). *International vocabulary of metrology. Basic and general concepts and associated terms (VIM)* (3rd ed.). Sèvres: Joint Committee for Guides in Metrology. http://www.bipm.org/en/publications/guides/vim.html.

Jóźwiak, G., Henrykowski, A., Masalska, A., & Gotszalk, T. (2012). Regularization mechanism in blind tip reconstruction procedure. *Ultramicroscopy, 118*, 1–10. http://dx.doi. org/10.1016/j.ultramic.2012.04.013.

Jóźwiak, G., Wielgoszewski, G., Gotszalk, T., & Kępiński, L. (2013). Thermal mapping of a scanning thermal microscopy tip. *Ultramicroscopy, 133*, 80–87. http://dx.doi.org/ 10.1016/j.ultramic.2013.06.020.

Ju, Y. (1999). Thermal characterization of anisotropic thin dielectric films using harmonic Joule heating. *Thin Solid Films, 339*(1–2), 160–164. http://dx.doi.org/10.1016/ S0040-6090(98)01328-5.

Juszczyk, J., Wojtol, M., & Bodzenta, J. (2013). DC experiments in quantitative scanning thermal microscopy. *International Journal of Thermophysics, 34*(4), 620–628. http://dx. doi.org/10.1007/s10765-013-1449-4.

Kaźmierczak-Bałata, A., Krzywiecki, M., Juszczyk, J., Firek, P., Szmidt, J., & Bodzenta, J. (2013). Application of scanning microscopy to study correlation between thermal properties and morphology of BaTiO$_3$ thin films. *Thin Solid Films, 545*, 217–221. http://dx. doi.org/10.1016/j.tsf.2013.08.007.

Kelvin Nanotechnology Ltd. (2015). Kelvin Nanotechnology – Products – Scanning Thermal Microscopy Probes. (online). http://www.kelvinnanotechnology.com/ products/products_scanning_probes.html.

Kim, K., Chung, J., Won, J., Kwon, O., Lee, J. S., Park, S. H., et al. (2008). Quantitative scanning thermal microscopy using double scan technique. *Applied Physics Letters, 93*(20), 203115. http://dx.doi.org/10.1063/1.3033545.

Kim, H. J., Dai, Z., & King, W. P. (2013). Thermal crosstalk in heated microcantilever arrays. *Journal of Micromechanics and Microengineering, 23*(2), 025001. http://dx.doi.org/10.1088/ 0960-1317/23/2/025001.

Kim, K., Jeong, W., Lee, W., & Reddy, P. (2012). Ultra-high vacuum scanning thermal microscopy for nanometer resolution quantitative thermometry. *ACS Nano, 6*, 4248–4257. http://dx.doi.org/10.1021/nn300774n.

Kim, K., Jeong, W., Lee, W., Sadat, S., Thompson, D., Meyhofer, E., et al. (2014). Quantification of thermal and contact resistances of scanning thermal probes. *Applied Physics Letters, 105*(20), 203107. http://dx.doi.org/10.1063/1.4902075.

Kim, K. J., & King, W. P. (2009). Thermal conduction between a heated microcantilever and a surrounding air environment. *Applied Thermal Engineering, 29*(8–9), 1631–1641. http:// dx.doi.org/10.1016/j.applthermaleng.2008.07.019.

Kim, S.-J., Ono, T., & Esashi, M. (2009). Thermal imaging with tapping mode using a bimetal oscillator formed at the end of a cantilever. *Review of Scientific Instruments, 80*(3), 033703. http://dx.doi.org/10.1063/1.3095680.

Klapetek, P. (2013). *Quantitative data processing in scanning probe microscopy.* Oxford, UK: William Andrew/Elsevier.

Lányi, Š. (2008). *Application of scanning capacitance microscopy to analysis at the nanoscale.* In B. Bhushan, H. Fuchs, & M. Tomitori (Eds.), *Applied scanning probe methods VIII. Nano science and technology* (pp. 377–420). Berlin and Heidelberg: Springer. http://dx.doi.org/10.1007/978-3-540-74080-3_11.

Lee, J.-H., & Gianchandani, Y. B. (2004). High-resolution scanning thermal probe with servocontrolled interface circuit for microcalorimetry and other applications. *Review of Scientific Instruments, 75*(5), 1222. http://dx.doi.org/10.1063/1.1711153.

Lee, J., & King, W. P. (2008). Improved all-silicon microcantilever heaters with integrated piezoresistive sensing. *Journal of Microelectromechanical Systems, 17*(2), 432–445. http://dx.doi.org/10.1109/JMEMS.2008.918423.

Lefèvre, S., Saulnier, J.-B., Fuentes, C., & Volz, S. (2004). Probe calibration of the scanning thermal microscope in the AC mode. *Superlattices and Microstructures, 35*(3–6), 283–288. http://dx.doi.org/10.1016/j.spmi.2003.11.004.

Lefèvre, S., & Volz, S. (2005). 3ω-scanning thermal microscope. *Review of Scientific Instruments, 76*(3), 033701. http://dx.doi.org/10.1063/1.1857151.

Lefèvre, S., Volz, S., & Chapuis, P.-O. (2006). Nanoscale heat transfer at contact between a hot tip and a substrate. *International Journal of Heat and Mass Transfer, 49*(1–2), 251–258. http://dx.doi.org/10.1016/j.ijheatmasstransfer.2005.07.010.

Lefèvre, S., Volz, S., Saulnier, J.-B., Fuentes, C., & Trannoy, N. (2003). Thermal conductivity calibration for hot wire based dc scanning thermal microscopy. *Review of Scientific Instruments, 74*(4), 2418. http://dx.doi.org/10.1063/1.1544078.

Linnemann, R., Gotszalk, T., Rangelow, I. W., Dumania, P., & Oesterschulze, E. (1996). Atomic force microscopy and lateral force microscopy using piezoresistive cantilevers. *Journal of Vacuum Science & Technology B: Microelectronics and Nanometer Structures, 14*(2), 856. http://dx.doi.org/10.1116/1.589161.

Luo, K., Majumdar, A., Herrick, R. W., & Petroff, P. (1997a). Scanning thermal microscopy of a vertical–cavity surface-emitting laser. *Applied Physics Letters, 71*(12), 1604. http://dx.doi.org/10.1063/1.119991.

Luo, K., Shi, Z., Varesi, J., & Majumdar, A. (1997b). Sensor nanofabrication, performance, and conduction mechanisms in scanning thermal microscopy. *Journal of Vacuum Science & Technology B: Microelectronics and Nanometer Structures, 15*(2), 349. http://dx.doi.org/10.1116/1.589319.

Majumdar, A., Carrejo, J. P., & Lai, J. (1993). Thermal imaging using the atomic force microscope. *Applied Physics Letters, 62*(20), 2501–2503. http://dx.doi.org/10.1063/1.109335.

Majumdar, A., Lai, J., Chandrachood, M., Nakabeppu, O., Wu, Y., & Shi, Z. (1995). Thermal imaging by atomic force microscopy using thermocouple cantilever probes. *Review of Scientific Instruments, 66*(6), 3584. http://dx.doi.org/10.1063/1.1145474.

Menges, F., Riel, H., Stemmer, A., Dimitrakopoulos, C., & Gotsmann, B. (2013). Thermal transport into graphene through nanoscopic contacts. *Physical Review Letters, 111*(20), 205901. http://dx.doi.org/10.1103/PhysRevLett.111.205901.

Menges, F., Riel, H., Stemmer, A., & Gotsmann, B. (2012). Quantitative thermometry of nanoscale hot spots. *Nano Letters, 12*(2), 596–601. http://dx.doi.org/10.1021/nl203169t.

Meyer, G., & Amer, N. M. (1988). Novel optical approach to atomic force microscopy. *Applied Physics Letters, 53*(12), 1045. http://dx.doi.org/10.1063/1.100061.

Meyer, G., & Amer, N. M. (1990). Simultaneous measurement of lateral and normal forces with an optical-beam-deflection atomic force microscope. *Applied Physics Letters, 57*(20), 2089. http://dx.doi.org/10.1063/1.103950.

Michels, T., & Rangelow, I. W. (2014). Review of scanning probe micromachining and its applications within nanoscience. *Microelectronic Engineering, 126*, 191–203. http://dx.doi.org/10.1016/j.mee.2014.02.011.

Mills, G., Zhou, H., Midha, A., Donaldson, L., & Weaver, J. M. R. (1998). Scanning thermal microscopy using batch fabricated thermocouple probes. *Applied Physics Letters*, *72*(22), 2900–2902. http://dx.doi.org/10.1063/1.121453.

Moczała, M., Sosa, N., Topol, A., & Gotszalk, T. (2014). Investigation of multi-junction solar cells using electrostatic force microscopy methods. *Ultramicroscopy*, *141*, 1–8. http://dx.doi.org/10.1016/j.ultramic.2014.02.007.

Murrell, M. P., Welland, M. E., O'Shea, S. J., Wong, T. M. H., Barnes, J. R., McKinnon, A. W., et al. (1993). Spatially resolved electrical measurements of $SiO_2$ gate oxides using atomic force microscopy. *Applied Physics Letters*, *62*(7), 786–788. http://dx.doi.org/10.1063/1.108579.

Nakabeppu, O., Chandrachood, M., Wu, Y., Lai, J., & Majumdar, A. (1995). Scanning thermal imaging microscopy using composite cantilever probes. *Applied Physics Letters*, *66*(6), 694–696. http://dx.doi.org/10.1063/1.114102.

Nanosurf, Inc. (2009). *Scanning thermal microscopy (SThM) with the easyScan 2 FlexAFM*. Application note, http://www.nanosurf.com/.

NanoWorld AG. (2015). AFM tip–CDT-FMR. http://www.nanoworld.com/pointprobe-conductive-diamond-coated-force-modulation-mode-afm-tip-cdt-fmr.

Ndieyira, J. W., Watari, M., Barrera, A. D., Zhou, D., Vögtli, M., Batchelor, M., et al. (2008). Nanomechanical detection of antibiotic-mucopeptide binding in a model for superbug drug resistance. *Nature Nanotechnology*, *3*(11), 691–696. http://dx.doi.org/10.1038/nnano.2008.275.

Nelson, B., & King, W. (2008). Modeling and simulation of the interface temperature between a heated silicon tip and a substrate. *Nanoscale and Microscale Thermophysical Engineering*, *12*(1), 98–115. http://dx.doi.org/10.1080/15567260701866769.

Nieradka, K., Gotszalk, T. P., & Schroeder, G. (2012). A novel method for simultaneous readout of static bending and multimode resonance-frequency of microcantilever-based biochemical sensors. *Sensors and Actuators B: Chemical*, *170*, 172–175. http://dx.doi.org/10.1016/j.snb.2011.05.032.

Nobelprize.org. (1986). Press release: The 1986 Nobel Prize in Physics. (online). http://www.nobelprize.org/nobel_prizes/physics/laureates/1986/press.html.

Nonnenmacher, M., O'Boyle, M. P., & Wickramasinghe, H. K. (1991). Kelvin probe force microscopy. *Applied Physics Letters*, *58*(25), 2921–2923. http://dx.doi.org/10.1063/1.105227.

Oesterschulze, E., & Stopka, M. (1996). Photothermal imaging by scanning thermal microscopy. *Journal of Vacuum Science & Technology, A: Vacuum, Surfaces, and Films*, *14*(3), 1172–1177. http://dx.doi.org/10.1116/1.580261.

Park Systems Corp. (2007). *XE Mode Note: Scanning thermal microscopy (SThM)*. Application note, Park Systems Inc. http://www.parkafm.com/AFM_guide/spm_modes_10.php?id=1190.

Park Systems Corp. (2010). *Scanning thermal microscopy (SThM) high spatial and thermal resolution microscopy by the XE-series innovations*. Application note, Park Systems Inc.

Pease, R. A. (2008). *A comprehensive study of the Howland current pump—AN-1515*. Application note, National Semiconductor.

Pollock, H. M., & Hammiche, A. (2001). Micro-thermal analysis: Techniques and applications. *Journal of Physics D: Applied Physics*, *34*(9), 2353. http://dx.doi.org/10.1088/0022-3727/34/9/201.

Preston-Thomas, H. (1990). The International Temperature Scale of 1990 (ITS-90). *Metrologia*, *27*(1), 3–10. http://dx.doi.org/10.1088/0026-1394/27/1/002.

Price, D., Reading, M., Hammiche, A., Pollock, H., & Branch, M. (1999). Localised thermal analysis of a packaging film. *Thermochimica Acta*, *332*(2), 143–149. http://dx.doi.org/10.1016/S0040-6031(99)00068-4.

Pumarol, M. E., Rosamond, M. C., Tovee, P., Petty, M. C., Zeze, D. A., Falko, V., et al. (2012). Direct nanoscale imaging of ballistic and diffusive thermal transport in graphene nanostructures. *Nano Letters*, *12*(6), 2906–2911. http://dx.doi.org/10.1021/nl3004946.

Puyoo, E., Grauby, S., Rampnoux, J.-M., Rouviere, E., & Dilhaire, S. (2011). Scanning thermal microscopy of individual silicon nanowires. *Journal of Applied Physics*, *109*(2), 024302. http://dx.doi.org/10.1063/1.3524223.

Rangelow, I. W., Gotszalk, T., Abedinov, N., Grabiec, P., & Edinger, K. (2001). Thermal nano-probe. *Microelectronic Engineering*, *57–58*, 737–748. http://dx.doi.org/10.1016/S0167-9317(01)00466-X.

Rugar, D., Mamin, H. J., & Güthner, P. (1989). Improved fiber-optic interferometer for atomic force microscopy. *Applied Physics Letters*, *55*(25), 2588–2590. http://dx.doi.org/10.1063/1.101987.

Ruiz, F., Sun, W. D., Pollak, F. H., & Venkatraman, C. (1998). Determination of the thermal conductivity of diamond-like nanocomposite films using a scanning thermal microscope. *Applied Physics Letters*, *73*(13), 1802–1804. http://dx.doi.org/10.1063/1.122287.

Sáenz, J. J., García, N., Grütter, P., Meyer, E., Heinzelmann, H., Wiesendanger, R., et al. (1987). Observation of magnetic forces by the atomic force microscope. *Journal of Applied Physics*, *62*(10), 4293. http://dx.doi.org/10.1063/1.339105.

Saïdi, E., Samson, B., Aigouy, L., Volz, S., Löw, P., Bergaud, C., et al. (2009). Scanning thermal imaging by near-field fluorescence spectroscopy. *Nanotechnology*, *20*(11), 115703. http://dx.doi.org/10.1088/0957-4484/20/11/115703.

Samson, B., Aigouy, L., Latempa, R., Tessier, G., Aprili, M., Mortier, M., et al. (2007). Scanning thermal imaging of an electrically excited aluminum microstripe. *Journal of Applied Physics*, *102*(2), 024305. http://dx.doi.org/10.1063/1.2756088.

Shi, L., Kwon, O., Majumdar, A., & Miner, A. (2001). Design and batch fabrication of probes for sub-100 nm scanning thermal microscopy. *Journal of Microelectromechanical Systems*, *10*(3), 370–378. http://dx.doi.org/10.1109/84.946785.

Shi, L., Kwon, O., Wu, G., & Majumdar, A. (2000). *Quantitative thermal probing of devices at sub-100 nm resolution*. In *Proceedings of the IEEE 38th annual international reliability physics symposium* (pp. 394–398), San Jose, CA: IEEE. http://dx.doi.org/10.1109/RELPHY.2000.843945.

Somnath, S., & King, W. P. (2013). Heated atomic force cantilever closed loop temperature control and application to high speed nanotopography imaging. *Sensors and Actuators A: Physical*, *192*, 27–33. http://dx.doi.org/10.1016/j.sna.2012.11.030.

STREP NANOHEAT. (2012). *Multidomain platform for integrated more-than-Moore/Beyond CMOS systems characterisation and diagnostics*. Project within the 7th Framework Programme of the European Union. http://www.nanoheat-project.org.

Sulzbach, T., & Rangelow, I. W. (Eds.). (2010). *PRONANO: Proceedings of the integrated project on massively parallel intelligent cantilever probe platforms for nanoscale analysis and synthesis*. Münster: Monsenstein und Vannerdat.

Terris, B., Stern, J., Rugar, D., & Mamin, H. (1989). Contact electrification using force microscopy. *Physical Review Letters*, *63*(24), 2669–2672. http://dx.doi.org/10.1103/PhysRevLett.63.2669.

Thiery, L., Gavignet, E., & Cretin, B. (2009). Two omega method for active thermocouple microscopy. *Review of Scientific Instruments*, *80*(3), 034901. http://dx.doi.org/10.1063/1.3097183.

Thiery, L., Marini, N., Prenel, J.-P., Spajer, M., Bainier, C., & Courjon, D. (2000). Temperature profile measurements of near-field optical microscopy fiber tips by means of sub-micronic thermocouple. *International Journal of Thermal Sciences*, *39*(4), 519–525. http://dx.doi.org/10.1016/S1290-0729(00)00231-3.

Thiery, L., Toullier, S., Teyssieux, D., & Briand, D. (2008). Thermal contact calibration between a thermocouple probe and a microhotplate. *Journal of Heat Transfer, 130*(9), 091601. http://dx.doi.org/10.1115/1.2943306.

Thomson, W. (1862). On the measurement of electric resistance. *Philosophical Magazine, 24*, 149–162. http://dx.doi.org/10.1080/14786446208643330.

Thomson Reuters. (2012). Web of Science—IP and Science. http://wokinfo.com/.

Tian, F., Qian, X., & Villarrubia, J. S. (2008). Blind estimation of general tip shape in AFM imaging. *Ultramicroscopy, 109*(1), 44–53. http://dx.doi.org/10.1016/j.ultramic.2008.08.002.

Tovee, P. D., & Kolosov, O. V. (2013). Mapping nanoscale thermal transfer in-liquid environment-immersion scanning thermal microscopy. *Nanotechnology, 24*(46), 465706. http://dx.doi.org/10.1088/0957-4484/24/46/465706.

Tovee, P., Pumarol, M., Zeze, D., Kjoller, K., & Kolosov, O. (2012). Nanoscale spatial resolution probes for scanning thermal microscopy of solid state materials. *Journal of Applied Physics, 112*(11), 114317. http://dx.doi.org/10.1063/1.4767923.

Vettiger, P., Cross, G., Despont, M., Drechsler, U., Durig, U., Gotsmann, B., et al. (2002). The "millipede" – nanotechnology entering data storage. *IEEE Transactions on Nanotechnology, 1*(1), 39–55. http://dx.doi.org/10.1109/TNANO.2002.1005425.

Villarrubia, J. (1994). Morphological estimation of tip geometry for scanned probe microscopy. *Surface Science, 321*(3), 287–300. http://dx.doi.org/10.1016/0039-6028(94)90194-5.

Villarrubia, J. (1997). Algorithms for scanned probe microscope image simulation, surface reconstruction, and tip estimation. *Journal of Research of the National Institute of Standards and Technology, 102*(4), 425–454. http://dx.doi.org/10.6028/jres.102.030.

Villarrubia, J. S., Scire, F., Teague, E. C., & Gadzuk, J. W. (2001). *The Topografiner: An instrument for measuring surface microtopography.* In D. R. Lide (Ed.), *A century of excellence in measurements, standards, and technology: A chronicle of selected NBS/NIST publications, 1901–2000* (pp. 214–218): National Institute of Standards and Technology.

Weaver, J. M. R., Walpita, L. M., & Wickramasinghe, H. K. (1989). Optical absorption microscopy and spectroscopy with nanometre resolution. *Nature, 342*(6251), 783–785. http://dx.doi.org/10.1038/342783a0.

Wheatstone, C. (1843). An account of several new instruments and processes for determining the constants of a voltaic circuit. *Philosophical Transactions of the Royal Society of London, 133*, 303–327. http://dx.doi.org/10.1098/rstl.1843.0014.

Wielgoszewski, G. (2014). *Metrologia właściwości termicznych mikro- i nanostruktur prowadzona metodami skaningowej mikroskopii termicznej bliskiego pola (Metrology of thermal properties of micro- and nanostructures using near-field scanning thermal microscopy).* Ph.D. thesis, Wrocław: Wrocław University of Technology.

Wielgoszewski, G., Babij, M., Szeloch, R. F., & Gotszalk, T. (2014a). Standard-based direct calibration method for scanning thermal microscopy nanoprobes. *Sensors and Actuators A: Physical, 214*, 1–6. http://dx.doi.org/10.1016/j.sna.2014.03.035.

Wielgoszewski, G., Gotszalk, T., Woszczyna, M., Zawierucha, P., & Zschech, E. (2008). Conductive atomic force microscope for investigation of thin-film gate insulators. *Bulletin of the Polish Academy of Sciences: Technical Sciences, 56*(1), 39–44. http://bulletin.pan.pl/(56–1)39.pdf.

Wielgoszewski, G., Jóźwiak, G., Babij, M., Baraniecki, T., Geer, R., & Gotszalk, T. (2014b). Investigation of thermal effects in through-silicon vias using scanning thermal microscopy. *Micron, 66*, 63–68. http://dx.doi.org/10.1016/j.micron.2014.05.008.

Wielgoszewski, G., Sulecki, P., Gotszalk, T., Janus, P., Grabiec, P., Hecker, M., et al. (2011a). Scanning thermal microscopy: A nanoprobe technique for studying the thermal properties of nanocomponents. *Physica Status Solidi B, 248*(2), 370–374. http://dx.doi.org/10.1002/pssb.201046614.

Wielgoszewski, G., Sulecki, P., Gotszalk, T., Janus, P., Szmigiel, D., Grabiec, P., et al. (2010). Microfabricated resistive high-sensitivity nanoprobe for scanning thermal microscopy. *Journal of Vacuum Science & Technology B: Microelectronics and Nanometer Structures, 28*(6), C6N7–C6N11. http://dx.doi.org/10.1116/1.3502614.

Wielgoszewski, G., Sulecki, P., Janus, P., Grabiec, P., Zschech, E., & Gotszalk, T. (2011b). A high-resolution measurement system for novel scanning thermal microscopy resistive nanoprobes. *Measurement Science and Technology, 22*(9), 094023. http://dx.doi.org/10.1088/0957-0233/22/9/094023.

Williams, C. C., & Wickramasinghe, H. K. (1986a). *High-resolution thermal microscopy.* In *IEEE 1986 Ultrasonics Symposium* (pp. 393–397). Williamsburg, VA: IEEE. http://dx.doi.org/10.1109/ULTSYM.1986.198771.

Williams, C. C., & Wickramasinghe, H. K. (1986b). Scanning thermal profiler. *Applied Physics Letters, 49*(23), 1587–1589. http://dx.doi.org/10.1063/1.97288.

Williams, C. C., & Wickramasinghe, H. K. (1990). Microscopy of chemical-potential variations on an atomic scale. *Nature, 344*(6264), 317–319. http://dx.doi.org/10.1038/344317a0.

Wollaston, W. H. (1813). A method of drawing extremely fine wires. Philosophical Transactions of the Royal Society of London, *103*, 114–118. http://dx.doi.org/10.1098/rstl.1813.0018.

Woszczyna, M., Zawierucha, P., Masalska, A., Jóźwiak, G., Staryga, E., & Gotszalk, T. (2010a). Tunneling/shear force microscopy using piezoelectric tuning forks for characterization of topography and local electric surface properties. *Ultramicroscopy, 110*(7), 877–880. http://dx.doi.org/10.1016/j.ultramic.2010.03.013.

Woszczyna, M., Zawierucha, P., Paşetko, P., Zielony, M., Gotszalk, T., Sarov, Y., et al. (2010b). Micromachined scanning proximal probes with integrated piezoresistive readout and bimetal actuator for high eigenmode operation. *Journal of Vacuum Science & Technology B: Microelectronics and Nanometer Structures, 28*(6), C6N12–C6N17. http://dx.doi.org/10.1116/1.3518465.

Yacoot, A., & Koenders, L. (2011). Recent developments in dimensional nanometrology using AFMs. *Measurement Science and Technology, 22*(12), 122001. http://dx.doi.org/10.1088/0957-0233/22/12/122001.

Yoon, K., Hwang, G., Chung, J., Kim, H. G., Kwon, O., Kihm, K. D., et al. (2014). Measuring the thermal conductivity of residue-free suspended graphene bridge using null point scanning thermal microscopy. *Carbon, 76*, 77–83. http://dx.doi.org/10.1016/j.carbon.2014.04.051.

Young, R. D. (1966). Field emission ultramicrometer. *Review of Scientific Instruments, 37*(3), 275–278. http://dx.doi.org/10.1063/1.1720157.

Young, R. D., Ward, J., & Scire, F. (1972). The Topografiner: An instrument for measuring surface microtopography. *Review of Scientific Instruments, 43*(7), 999–1011. http://dx.doi.org/10.1063/1.1685846.

Yu, Y.-J., Han, M. Y., Berciaud, S., Georgescu, A. B., Heinz, T. F., Brus, L. E., et al. (2011). High-resolution spatial mapping of the temperature distribution of a Joule self-heated graphene nanoribbon. *Applied Physics Letters, 99*(18), 183105. http://dx.doi.org/10.1063/1.365751.

Zhang, Y., Dobson, P. S., & Weaver, J. M. R. (2011). Batch fabricated dual cantilever resistive probe for scanning thermal microscopy. *Microelectronic Engineering, 88*(8), 2435–2438. http://dx.doi.org/10.1016/j.mee.2011.02.040.

Zhang, Y., Dobson, P. S., & Weaver, J. M. R. (2012). High-temperature imaging using a thermally compensated cantilever resistive probe for scanning thermal microscopy. *Journal of Vacuum Science & Technology B: Microelectronics and Nanometer Structures, 30*(1), 010601. http://dx.doi.org/10.1116/1.3664328.

# CONTENTS OF VOLUMES 151-189

---

[1] Lists of the contents of volumes 100–149 are to be found in volume 150; the entire series can be
searched on ScienceDirect.com

# INDEX

Note: Page numbers followed by "*f*" indicate figures and "*t*" indicate tables.

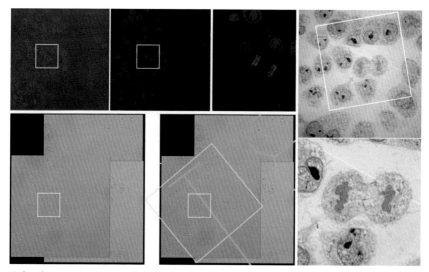

**Niels de Jonge, Figure 1** CLEM allows the identification and subsequent high-resolution analysis of 1 special cell among hundreds. In the upper-left row, a dividing cell (1 in >100) is identified based on its DNA staining (blue). With the aid of embossed coverslips (for full details, see Hodgson *et al.*, 2014a) the dividing cell can be traced back in the EM and studied at higher resolutions.

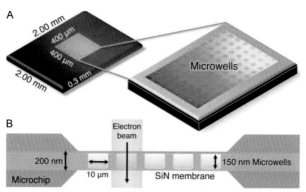

**Niels de Jonge, Figure 3** Next-generation SiN microchips. (A) A schematic to specify the dimensions of the microwell chips used to form the liquid chamber that is positioned with respect to the electron beam **(B)**. *Illustrations adapted and reprinted with permission (Dukes* et al., *2014).*

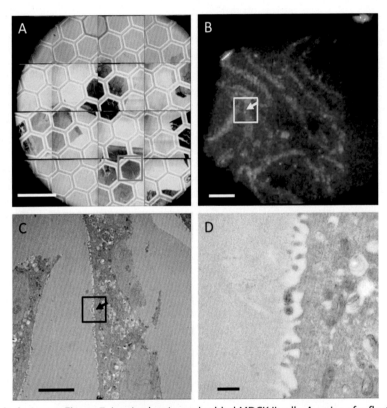

**Niels de Jonge, Figure 7** Lowicryl resin–embedded MDCK II cells. A series of reflection images (A; scale bar 250 $\mu$m) allows us to localize the sample. Immuno-fluorescence (B; scale bar 25 $\mu$m) indicates the locations of acetylated alpha-tubulin as found in cilia. Finally, we obtain TEM images (C; scale bar 5 $\mu$m) and (D; scale bar 500 nm) of the regions of interest found by fluorescence.

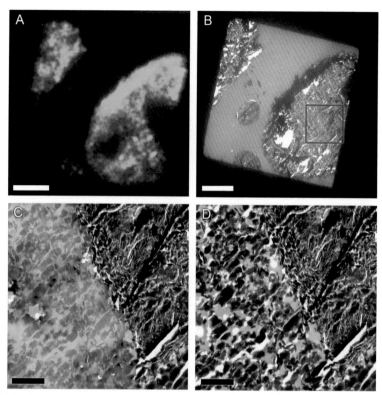

**Niels de Jonge, Figure 8** iLEM analysis of a fluid catalyst cracking (FCC) particle. Active regions in the particle generate a fluorescent product (A; scale bar 10 $\mu$m). A zoom, indicated by a blue box, into the TEM image (B; scale bar 10 $\mu$m) shows that fluorescence and specific morphology are correlated (C, D; scale bar 2 $\mu$m).

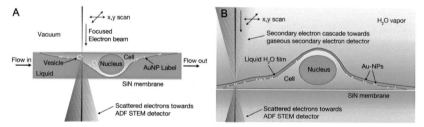

**Niels de Jonge, Figure 20** Principles of Liquid STEM. A whole cell containing proteins labeled with gold nanoparticles (AuNPs) is imaged using the annular dark field (ADF) detector beneath the sample. (A) The cell is fully enclosed in a microfluidic chamber with two SiN windows for STEM. (B) The cell is maintained in a saturated water vapor atmosphere, while a thin layer of water covers the cell for ESEM-STEM. *Used with permission from Peckys & de Jonge (2014).*

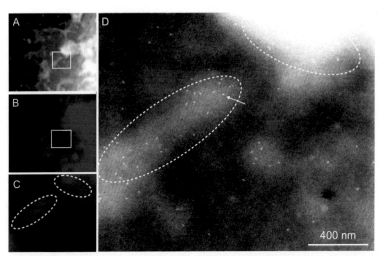

**Niels de Jonge, Figure 21** Correlative fluorescence microscopy and ESEM-STEM of quantum dot labeled ErbB2 receptors on a SKBR3 cell. (A) Overview of ESEM-STEM image. (B) Fluorescence image recorded from the same area as shown in (A). (C) Enlarged detail from the image shown in (C) of the same size as (D). (D) STEM image from the boxed region shown in (A). The locations of individual HER2 receptors are indicated by the bright, bullet-shaped QDs (example at arrow). HER2 concentrates on electron denser (bright) areas (enclosed in dashed lines); i.e., membrane ruffles.

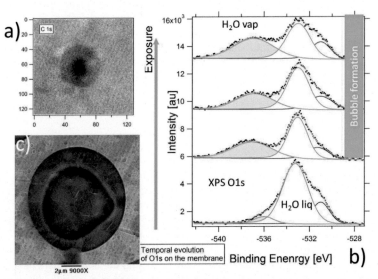

**Niels de Jonge, Figure 22** (left) TPP_2M calculated inelastic electron mean free path in carbon. The thickness of one and two layers of graphene is depicted in the bottom of the graph. Inset: E-cell principle design of graphene E-cell and electron attenuation formula; (right) the comparative SEM image of a water-filled GE cell, with and without Au nanoparticles.

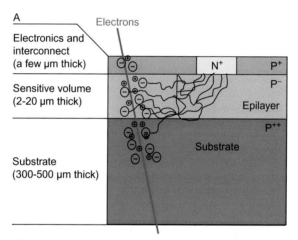

**A.R. Faruqi et al., Figure 1** (A) Typical side view of a 3T pixel in a MAPS. The incident electron leaves a trail of electron–hole pairs along the track. Electrons created within the epilayer have a sufficiently long mean free path to diffuse to the N$^+$ diode, buried in the P$^+$ well, helped by a slight potential difference due to different doping densities in P$^-$ and N$^+$ silicon. The substrate consists of highly doped silicon, which acts as an inert structure with rapid electron–hole recombination, so that it plays only a marginal role in signal generation (Turchetta et al., 2001).

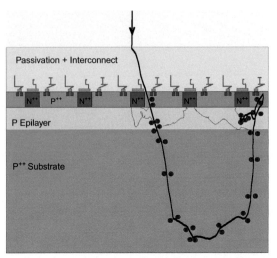

**A.R. Faruqi *et al.*, Figure 2** Monte Carlo simulation of a single-electron trajectory through a pixel, which was backscattered in the silicon substrate. The passage of the electron creates electron–hole pairs along the track, and the electrons and holes are shown as blue and red circles, respectively. Electrons created in the epilayer diffuse to the collecting diode. Toward the end of the track in the backscattered electron, there is a much higher concentration of electron–hole pairs as the electron has much lower energy, resulting in a far higher signal that is also in the wrong place (McMullan, Chen, Henderson, & Faruqi, 2009; McMullan et al., 2009).

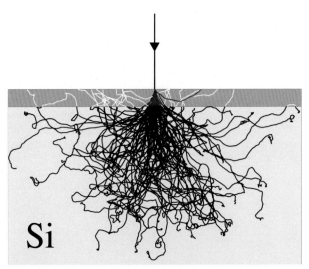

**A.R. Faruqi *et al.*, Figure 3** Monte Carlo simulation showing trajectories of 300-keV electrons through a silicon sensor with a thickness of 350 $\mu$m with the top part (shown in dark gray) being 35 $\mu$m. The incident electron is shown in black. Electrons giving a high DQE are shown in red, apart from one electron, which is backscattered in the top layer. The white tracks show backscattered electrons that create a signal away from the impact position and decrease the DQE at high spatial frequencies (McMullan et al., 2009). This simulation suggests that a back-thinned sensor with a thickness of 35 $\mu$m should have superior DQE, and this has been confirmed (McMullan et al., 2009).

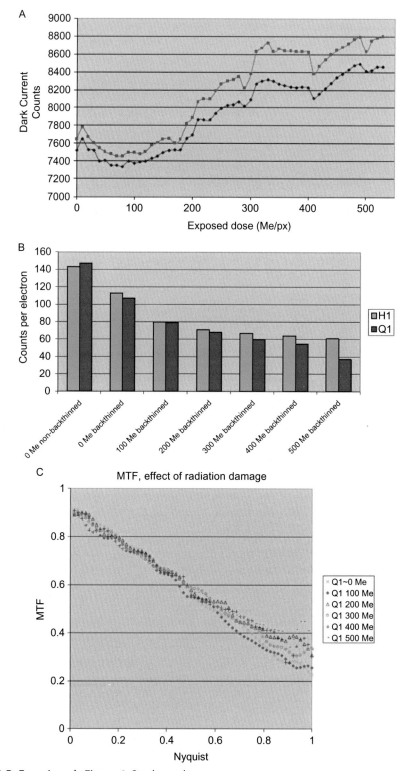

A.R. Faruqi *et al.*, **Figure 4** See legend on next page.

**A.R. Faruqi** *et al.*, **Figure 5** The commercial version of the 4k × 4k detector, Falcon (http://www.fei.com/), ready for mounting in the electron microscope.

**Figure 4** (A) Variation of dark current as a function of irradiation in two chosen subareas as a function of dose in units of million electrons/pixel (Guerrini et al., 2011a). The blue data points come from the more rad-hard pixel architecture. (B) Variation in sensitivity in units of counts per 300-keV electron for the two chosen pixels as a function of irradiation level. The first data point at 0 Me is for a non-back-thinned sensor, whereas all the other points are for a back-thinned sensor; (Guerrini et al., 2011a). The blue data bars show data from the more rad-hard pixel architecture. (C) Plots showing the variation of MTF as a function of spatial frequency (up to Nyquist frequency) for different levels of irradiation for one chosen pixel architecture (Guerrini et al., 2011a).

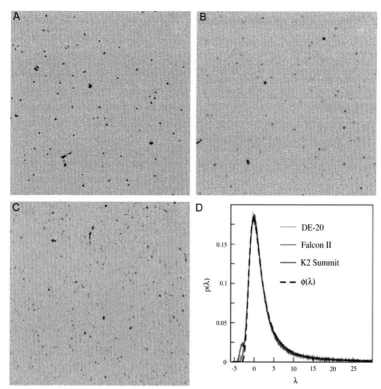

**A.R. Faruqi _et al._, Figure 6** Single-electron events taken from 256 x 256 pixel areas at 300 keV from three detectors: (A) DE-20, (B) Falcon II, and (C) K2 Summit. The Landau plots for all three detectors are shown in (D), along with the expected normalized Landau plot. For details, see McMullan et al. (2014).

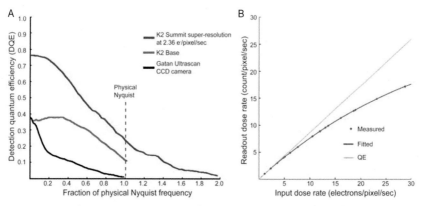

**A.R. Faruqi _et al._, Figure 8** (A) DQE as a function of spatial frequency for K2 Summit in linear (analog) mode, counting mode, and comparison with an Ultrascan CCD. (B) Relation between input and output rates, which have a linear relation up to 5 electrons/pixel/s, but some losses occur at higher counting rates (Li et al., 2013). Used with permission from Nature Methods.

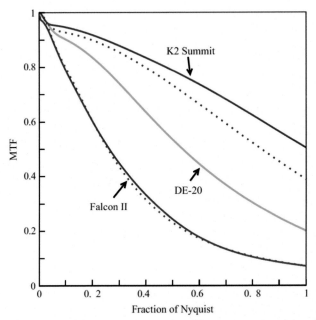

**A.R. Faruqi** *et al.*, **Figure 9** Experimental plots for MTF for the three detectors. DE-20 is shown in green; K2 Summit shown in blue, with super-resolution counting mode (shown as a line) and normal counting mode (shown as dots); Falcon II is shown as red line with derived MTF from NPS shown as dots (McMullan et al., 2014).

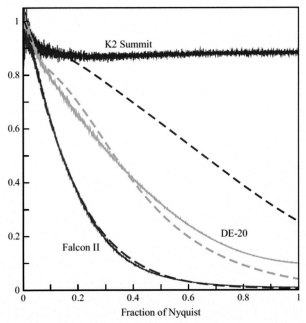

**A.R. Faruqi** *et al.*, **Figure 10** Noise power spectrum as a function of spatial frequency for K2 Summit shown in blue, DE20 shown in green, and Falcon II shown in red. Dashed lines represent $MTF^2$ measurements (McMullan et al., 2014).

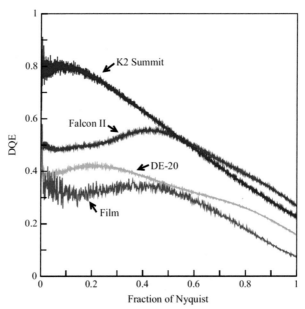

**A.R. Faruqi *et al.*, Figure 11** Experimental measurements of DQE as a function of spatial frequency for K2 Summit in super-resolution; counting mode in blue, Falcon II in red, DE-20 in green, and film in black (McMullan et al., 2014).

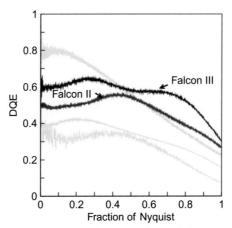

**A.R. Faruqi *et al.*, Figure 12** DQE as a function of spatial frequency at 300 keV for the back-thinned Falcon II with a sensor thickness of about 50 $\mu$m, and the further back-thinned Falcon III with a thickness of about 30 $\mu$m, showing improved DQE at all spatial frequencies.

**A.R. Faruqi *et al.*, Figure 15** Image of single particles of β-galactosidase illustrating the tracking algorithm used by Vinothkumar, McMullan, and Henderson (2014). The dotted circle indicates the region from which a group of particles (square box) was used for tracking.

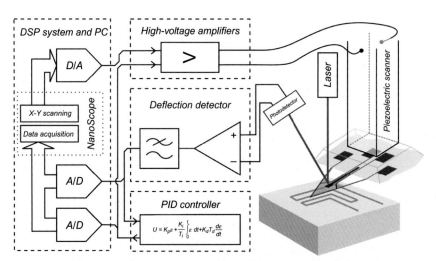

**Grzegorz Wielgoszewski and Teodor Gotszalk, Figure 2** Schematic of an AFM measurement system with optical detection of the deflection of the cantilever, in which it is the microprobe that is mounted on the piezoelectric scanner (see also Wielgoszewski et al., 2011b).

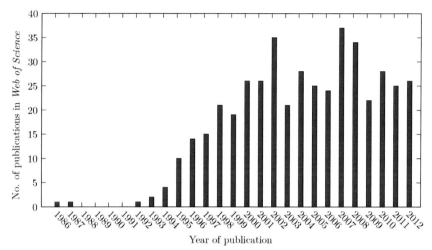

**Grzegorz Wielgoszewski and Teodor Gotszalk, Figure 3** Number of publications on SThM released in 1986–2012, based on Thomson Reuters Web of Science (Wielgoszewski, 2014).

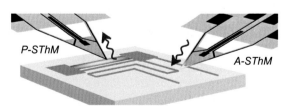

**Grzegorz Wielgoszewski and Teodor Gotszalk, Figure 5** A simple differentiation between SThM operating modes: passive (P-SThM) and active (A-SThM).

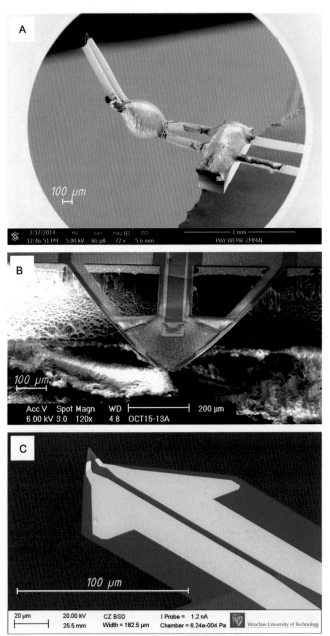

**Grzegorz Wielgoszewski and Teodor Gotszalk, Figure 9** Examples of thermoresistive nanoprobes: (A) Wollaston-wire probe, (B) ITE/WRUT nanoprobe, (C) KNT-SThM nanoprobe. Blue—platinum/palladium; light blue—silver; yellow—gold; green—$Si_3N_4$; pink—mounting glue; purple—glass. SEM images courtesy of Karolina Orłowska (Wrocław University of Technology), Yvonne Ritz (AMD Saxony), and Patrycja Szymczyk (Wrocław University of Technology).

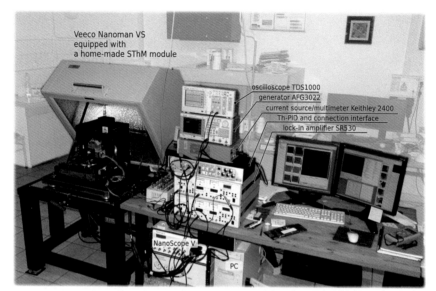

**Grzegorz Wielgoszewski and Teodor Gotszalk, Figure 13** An overview of a commercial AFM equipped with a homemade SThM system, as used in the Nanometrology Group at Wrocław University of Technology.

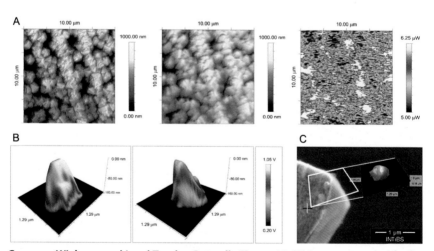

**Grzegorz Wielgoszewski and Teodor Gotszalk, Figure 14** (A) Images recorded for further processing in TThM: the reference high-resolution AFM topography (left) and simultaneously recorded SThM topography and T-ASThM heat dissipation from the tip to the sample; (B) the estimated tip shape with the thermal map seen as color overlay; (C) SEM image as a confirmation of the correct tip shape estimation. High-resolution SEM image courtesy of Leszek Kępiński (Institute of Low Temperatures and Structure Research of the Polish Academy of Sciences); blue—platinum; green—$SiO_2$. *Reprinted from Jóźwiak et al. (2013), with permission from Elsevier.*

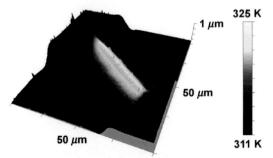

**Grzegorz Wielgoszewski and Teodor Gotszalk, Figure 15** Passive-mode SThM image of a microfuse made of polycrystalline silicon, obtained using a calibrated tip while a 10-mA current was flowing through the structure; the temperature map is overlaid on the recorded surface topography. *Reprinted from Wielgoszewski et al. (2014), with permission from Elsevier.*

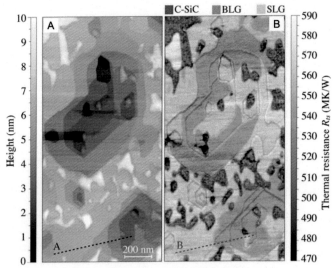

**Grzegorz Wielgoszewski and Teodor Gotszalk, Figure 16** (A) Topography and (B) active-mode SThM image of a SiC sample with single-layer graphene (SLG) and bilayer graphene (BLG) formed on the surface. *Reprinted with permission from Menges et al. (2013) © by the American Physical Society.*

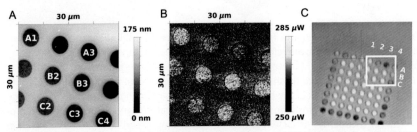

**Grzegorz Wielgoszewski and Teodor Gotszalk, Figure 17** (A) Topography and (B) tip power images recorded in constant-temperature active-mode SThM, accompanied by (C) optical image of the structure. *Reprinted from Wielgoszewski et al. (2014), with permission from Elsevier.*

Printed in the United States
By Bookmasters